Site safety handbook

Third Edition

prepared by

S C Bielby

revised by

J A Read

sharing knowledge ■ *building best practice*

6 Storey's Gate, Westminster, London SW1P 3AU
TELEPHONE 020 7222 8891 Fax 020 7222 1708
EMAIL enquiries@ciria.org.uk
WEBSITE www.ciria.org.uk

> **A half of all the people who die on a construction site have been working there for less than two weeks. Effective management action could have prevented 70 per cent of fatal construction accidents.**

A hazard is something with the potential to cause injury, eg an unsupported trench 1.8 m deep.

Risk is the likelihood that the hazard will result in an accident. Risk also considers how serious the resultant injury would be.

Construction sites are hazardous. These hazards should have been the subject of an assessment by your employer, and/or those who created the hazard, to assess the risks and put in place arrangements to eliminate or control them.

Nevertheless your health and your safety should be uppermost in your mind while you are working on site to satisfy yourself that those arrangements are adequate and effective and address the many hazards you will encounter on a day-to-day basis.

Kevin Myers
Chief Inspector of Construction
Health and Safety Executive

You should also be thinking of your colleagues and others working on site, as well as members of the general public, whose the health and safety might be affected by your construction operations.

This easy-to-use reference guide gives an overview to prepare you for working safely, advice on what to do when you come across commonly encountered hazards or if an accident occurs.

Everyone on site should find it helpful and useful. It also takes account of recent changes in health and safety legislation, for example *The Lifting Operations and Lifting Equipment Regulations 1998*, and emphasises the need for all those involved in construction to identify, assess and manage risks. The bibliography of sources of further information has also been extensively revised in this edition.

I am pleased that CIRIA has produced this revised handbook for the benefit of all those in the construction industry. I commend it to you and urge you to read it and follow the advice it contains.

Summary

Building and civil engineering construction sites contain many hazards. The risks to engineers, architects and surveyors – and especially to young professionals going to work on them for the first time – are considerable.

This third edition of CIRIA's Site safety handbook has been prepared to alert construction professionals to the hazards often present on site, to provide advice on safer practices for themselves and others, and to help them manage these important responsibilities. It has been revised to reflect the recent changes in health and safety regulations.

Although the text is written in the context of the legal framework provided by the Health and Safety at Work etc. Act of 1974, the handbook is not a legal document nor is it a comprehensive manual on site safety. It is designed to be an easy-to-read, ready reference guide for use outside, which will slip easily into a pocket. To achieve its purpose, the text has been kept brief.

The text is divided into four main parts:

- **Before going on site** – four chapters explain the responsibilities of the young professional and the preparations needed before going on site
- **Site activities and hazards** – each of the 27 chapters deals with a principal risk, eg falling, temporary works
- **Your contribution** – three chapters explain what professionals should do to deal with hazards, assist with accidents and how to investigate them
- **Further help and information** – contains a reference section and an index.

The handbook starts with a foreword by *Kevin Myers, Chief Inspector, Health and Safety Executive*

Summary

Site safety handbook (third edition).

Bielby, S C and Read, J A
Special Publication 151
First published as SP90 in 1992; second edition (as SP 130) 1997, reprinted in paperback 2000 (twice); third edition published June 2001, reprinted November 2001.

CIRIA SP151   © CIRIA 2001   ISBN 0 86017 800 5

Keywords
Construction site, health and safety, risks, risks assessment, hazards, law, young professional, personal preparation, protective equipment, accidents, working practices, training, safety legislation, management, precautions

Reader interest	Classification	
Construction professionals – engineers, architects, surveyors, planners; clients; consultants; local authorities; safety officers; students; site managers.	AVAILABILITY CONTENT STATUS USER	Unrestricted Advice/guidance Committee-guided Construction professionals and managers

Contents

Contents

Acknowledgements

The revision of CIRIA's *Site safety handbook* for this third edition was carried out by Mr J A Read of WS Atkins.

The technical updating for the third edition of *Site safety handbook* has been funded by CIRIA who wishes to express its thanks to all who contributed to this and previous editions of the handbook:

CIRIA's research manager for the project was D W Churcher.

CIRIA is grateful to the following organisations which made material available:

Ove Arup Partnership	Health and Safety Executive
WS Atkins Ltd	HMSO
Balfour Beatty Ltd	Institution of Civil Engineers
Building Employers Confederation	Royal Institution of Chartered Surveyors
Construction Industry Training Board	Thames Water Utilities
Costain Group	Travers Morgan Ltd.
HMSO	RoSPA
V J Davies	CITB
Tiefbau Berufsgenossenschaft	

Note

Recent Government reorganisation has meant that DETR responsibilities have been moved variously to the Department of Trade and Industry (DTI), the Department for the Environment, Food and Rural Affairs (DEFRA), and the Department for Transport, Local Government and the Regions (DTLR). References made to the DETR in this publication should be read in this context.

For clarification, readers should contact the Department of Trade and Industry.

Starting point

Construction professionals

Did you realise that every fourth working day (on average) someone dies on a British construction site? And that is something that legislation alone cannot change. This handbook is essential reading when you go on site, regardless of your job description. It is directly relevant to all construction professionals, including:

- civil engineers
- structural engineers
- architects
- quantity surveyors
- building services engineers
- project managers

- planners
- electrical engineers
- mechanical engineers
- building engineers
- surveyors
- facilities managers.

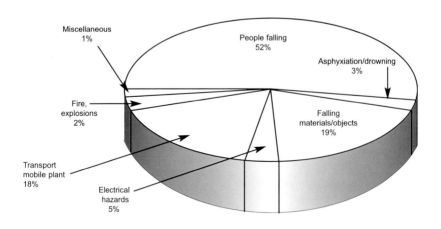

The major causes of construction deaths

 You have a 1 in 2 chance of injury during 20 years working on site.

Scope of the handbook

This handbook has been designed as a ready reference guide to advise you about health and safety and to help you understand and manage health and safety in construction. *However, it is not a legal document nor is it a comprehensive manual.*

It highlights the common hazards that you will be meeting from day to day. It promotes safer construction practices by explaining the basic safety standards you will need to apply on construction sites. The handbook does not cover ionising radiation, compressed air, explosives, quarrying, diving, tunnelling or other specialist activities. Specialist training is required for these activities before you go on such sites.

The handbook is in four main sections:

- Before going on site
- Site activities and hazards
- Your contribution
- Further help and information.

The section on "Site activities and hazards" consists of 27 brief chapters arranged in alphabetical order. The figures illustrate basic safety points, but are not definitive working drawings.

It slips easily into your pocket so that it is at hand wherever and whenever you need it. This book could help save your life ... or the life of one of your colleagues.

This handbook has been written for you... because safety on site is something you cannot afford to ignore, and **you** have statutory responsibilities for safety.

1.1 Your responsibilities

You must work within this framework to manage health and safety:

Health and safety is integral to the responsibilities of everyone on site and must be exercised within a management framework. This includes:

- legal requirements
- your own company policy and its implementation
- specific site rules for safe systems of work
- professional codes of conduct and ethics.

As a professional, you can play an important part in helping to encourage safer construction practices and to prevent dangerous acts that could lead to accidents.

 Within the European Union (EU) there are 8000 industrial deaths (including the construction industry) each year.

The law

If there is a conflict between health and safety legislation, contract requirements, company policy, site rules or professional ethics, then health and safety legislation takes precedence.

The Health and Safety at Work etc Act 1974 provides a comprehensive legislative framework for promoting, stimulating and encouraging high standards of health and safety in the workplace. The aim of the Act is to increase health and safety awareness and to promote effective standards in every organisation.

Everyone is involved: management, employees, self-employed, employees' representatives, those in charge of premises and the manufacturers of plant, equipment, substances and materials in matters of health and safety. The Act also deals with the protection of the public, where they may be affected by work activities.

The Act consists of four parts, but only Part one directly concerns you.

The main provisions of Part one of the Act seek:

- to ensure the health, safety and welfare of persons at work
- to protect persons other than persons at work against risks to health and safety arising out of, or in connection with, the activities of persons at work
- to control the keeping and use of explosive or highly flammable or otherwise dangerous substances
- to control the emission into the atmosphere of noxious or offensive substances
- to provide for criminal offences and punishments.

The Act established a Health and Safety Commission (HSC) and Health and Safety Executive (HSE) to enforce and administer the Act and the Regulations made under it.

1.3 Your responsibilities

The Act says that your responsibilities are:

a) to take reasonable care for the health and safety of yourself and others who may be affected by your acts or omissions at work

b) to co-operate with your employer and any other person properly involved in all matters relating to health and safety law and any duty or requirement that the employer may be required to make under that law.

It is also an offence for any person intentionally or recklessly to interfere with or misuse anything provided in the interests of health, safety or welfare that the law may require.

You will also need to understand other Acts and Regulations as your career progresses, including:

- The Confined Space Regulations 1997
- The Construction (Design and Management) Regulations 1994 (CDM)
- The Construction (Head Protection) Regulations 1989
- The Construction (Health, Safety and Welfare) Regulations 1996 (CHSW)
- The Control of Substances Hazardous to Health Regulations 1999 (COSHH)
- The Electricity at Work Regulations 1989
- The Fire Precautions (Workplace) Regulations 1997
- The Lifting Operations and Lifting Equipment Regulations 1998 (LOLER)
- The Management of Health and Safety at Work Regulations 1999
- The Manual Handling Operations Regulations 1992
- The Noise at Work Regulations 1989
- The Provision and Use of Work Equipment Regulations 1998.
- Reporting of Injuries, Diseases and Dangerous Occurences Regulations 1995 (RIDDOR)

Consult the Construction Safety Manual for a complete list of Acts and Regulations relevant to construction. Note that these will be amended from time to time.

Company policy and its implementation

Every employer with five or more employees must provide a written statement of the company's general policy, organisation and arrangements for health and safety at work.
The employers must show, or give this, to all employees and to keep it up to date.

A statement of company safety policy will explain:

- what your employer intends should happen
- how the employer is going to set up and maintain a safe and healthy working environment
- what health and safety responsibilities exist
- who is responsible and how to contact them
- that safe systems of work exist and who is responsible for them
- the arrangements for review and update etc.

You must read this policy, understand your responsibilities and carry them out.

It is important that you know the organisational structure and whom you should ask for health and safety advice. Normally this will be your manager or the appointed safety adviser/supervisor for your site.

Risk assessment

The Management of Health and Safety at Work Regulations require that the risks associated with any hazardous work activity are assessed before work starts so that the necessary preventative and protective measures can be identified and put into place. This process of risk assessment starts at the design and planning stage of a project and continues during the construction phase. Risk assessment forms the basis of all recent health and safety legislation and regulations, and is the starting point for all design and safe systems of work.

Always ask to see the risk assessment for any potentially hazardous activity before you start work on that activity.

1.5 Your responsibilities

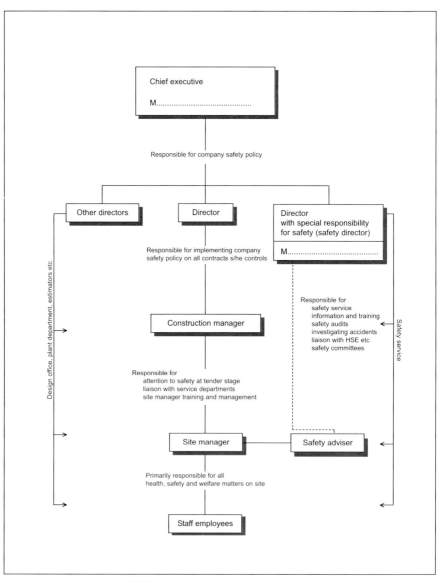

Responsibilities for health and safety in a typical company

Site rules for safe systems of work

Every site is unique and requires specific rules for safe systems of working. These are normally contained in written method statements prepared after hazards have been identified and risks assessed. They are then incorporated in the health and safety plan, which should cover all hazardous operations, including:

- frame erecting
- roofing
- cladding
- painting at height
- demolition
- hot working
- scaffolding
- working in confined spaces
- removing asbestos
- excavating
- pressure testing
- working over water.

Note! Following a risk assessment prepared for a hazardous activity, a method statement (where necessary) should be prepared.

The order of preparation should be:

1. Identify hazards
2. Assess risks (identify who is at risk)
3. Reduce risks to acceptable level using the principles of prevention (See 2.4)
4. Specify protective measures
5. Prepare method statement.

A method statement will typically include information on:

- the person in charge of operations
- safe means of access to and from all work areas
- specific details of any lifting appliances, including positioning on site and lifting gear to be used
- details of storage of materials and methods for dealing with hazardous substances
- how the work is to be carried out
- communications
- the equipment to be used
- protective clothing and equipment to be worn.

No deviation from the method statement should be allowed without referring back to the main contractor, a revised statement being produced and the health and safety plan being updated.

Risk assessments should always come before method statements. They should be explained to the people carrying out the work.

1.7 Your responsibilities

Professional codes of conduct

All members of professional bodies are bound by their code of conduct. For example, item 1 of the Rules for Professional Conduct of the ICE states that:

"A member, in his responsibility to his employer and to the profession, shall have full regard to the public interest, particularly in matters of health and safety."

Other institutions have similar rules.

Your actions

If you see a situation that, in your opinion, involves a risk of death or serious injury, you have a **statutory duty** to take immediate action.

This includes one or more of the following actions:

- tell the person in danger to stop, explaining why – but don't put yourself at risk in doing so
- contact the person in charge of the activity and your immediate manager and inform them of your actions
- later inform appropriate senior site personnel
- if you are uncertain about the degree of risk involved, consult the senior site manager before taking any further action and record it in your site diary.

If you ignore a danger you condone it and will share responsibility for any accident. Do not be put off or let yourself be overridden if you believe the danger exists. Things that look unsafe usually are.

Other risks should be dealt with through your formal channels of communication – refer the matter to your manager and your safety adviser. The HSE provides advice and guidance on health and safety matters. Consult it if you cannot get advice within your own organisation.

Never be frightened to ask to see a risk assessment if you believe the activity to be hazardous.

....

....

....

....

....

....

....

....

....

....

....

....

....

....

....

....

....

....

....

....

REMEMBER
**Firms *and* individuals can be prosecuted. If in doubt, ask
your immediate manager without delay.**

2.1 Construction-related regulations

The Construction (Design and Management) Regulations

When does CDM apply?

The CDM Regulations apply to construction projects that are notifiable to the HSE, ie those that will continue for more than 30 days or will involve more than 500 person-days of work. CDM also applies to non-notifiable work involving five people or more on site at any one time, and to all work that involves demolition.

CDM may not apply in full to work for certain clients, eg domestic clients. Further guidance on the application of CDM is given in *A guide to managing health and safety in construction* or the Approved Code of Practice.

The term "construction work" has been widely defined in the CDM Regulations and includes many activities within the building, civil engineering or engineering construction sectors of the construction industry.

New duties

The CDM Regulations place duties and responsibilities on clients, planning supervisors, designers, principal contractors and contractors, to plan, co-ordinate and manage health and safety throughout all stages of a construction project.

A health and safety plan has to be prepared and developed throughout the project.

REMEMBER Consult the health and safety plan.

Main duties of the five key parties

- **The client** – shall appoint a competent planning supervisor and principal contractor for each project.

- **The designer** – should design structures to avoid, wherever possible, foreseeable risks to health and safety during construction, maintenance and cleaning work. Information should be provided on unavoidable risks. Design includes the preparation of specifications – it is not limited to calculations and drawings.

- **The planning supervisor** – the organisation or person with overall responsibility for ensuring co-ordination of the health and safety aspects of the design and planning phase, the early stages of the health and safety plan and the health and safety file.

- **The principal contractor** – develops the construction-phase health and safety plan and manages and co-ordinates the activities of all contractors to ensure they comply with health and safety legislation. They have duties relating to the provision of information and training on health and safety for everyone on site and the co-ordination of employees' views working on the site.

- **Contractors and the self-employed** – must co-operate with the principal contractor and other contractors and provide relevant information on the risks created by their work and how they will be controlled.

Anyone arranging for anyone else to do work should make sure those they appoint are competent and sufficiently resourced to discharge their health and safety duties on that project.

Reg 13 duties on designers apply to all design work.

2.3 Construction-related regulations

The Construction (Health, Safety and Welfare) Regulations

These Regulations update and amend legislation originally published in the 1960s. The main duty-holders under these Regulations are employers, the self-employed **and those who control** the way in which construction work is carried out.

Contents

The Regulations deal with many specific aspects of site safety including:

- prevention of falls and falling materials
- working on fragile roofs
- excavation
- scaffolds
- demolition
- control of site traffic
- segregation of pedestrians and site traffic
- fire prevention, alarm and detection, means of escape and fire-fighting
- emergency procedures, eg flooding
- welfare facilities
- training
- inspection and reports.

The requirements of these Regulations have been reflected in this handbook.

The Management of Health and Safety at Work Regulations

These Regulations apply to all work, including construction. They include the requirement for employers (and the self-employed) to assess the risks arising from work activities.

The five steps to risk assessments

Step 1: Look for the hazards

Step 2: Decide who might be harmed and how

Step 3: Evaluate the risks and decide whether the existing precautions are adequate or whether more should be done

Step 4: Record your findings

Step 5: Review your assessment and revise it if necessary

In selecting control methods, the hierarchy of prevention and protection should be used:

- avoid risk altogether if possible
- combat risk at source
- wherever possible adapt work to the individual
- take advantage of technological progress
- ensure risk controls are part of a coherent policy
- protect the whole workforce rather than individuals
- ensure the control measures are understood.

Workers must be adequately trained to understand and implement the control measures. Employers must monitor the workplace to ensure that the control measures are implemented and effective; this will include appropriate health surveillance, eg lung function tests for spray painters.

REMEMBER
The use of Personal Protective Equipment (PPE) should be the last – not the first – choice of risk control.

3.1 Getting ready

Think ahead – what will you be doing on site?
Then:

- find out about relevant legislation and the standards it requires you to follow
- check your company policy statement and local arrangements and think how you will comply with them
- on projects to which the CDM Regulations apply, read the pre-tender health and safety plan
- read relevant risk assessments and method statements
- obtain the necessary equipment, protective clothing and materials
- make sure you understand company working practices
- evaluate your need for further training.

On joining a new site, you should:

- report to the site office and read the site notices
- discuss your previous safety training/experience (if any) with your immediate manager
- read the company safety policy and have the safety organisation explained
- be given induction training, including site safety rules and risk assessments and method statements, and receive instructions in safety procedures
- be told about the construction-phase health and safety plan (on projects to which the CDM Regulations apply)
- be given the name and location of the safety adviser
- be informed of the need to use protective clothing and equipment
- bring with you your safety equipment
- be told of your personal responsibilities for health and safety
- have explained to you the authorisation necessary for the use of plant, machinery, powered hand tools etc
- be told how to report "near misses" and defective plant and equipment

See also Chapter 4

- see the first aid facilities
- see the site and discuss its hazards.

Discuss these points with your manager or safety adviser to ensure you understand them.

You will also need to inform your manager of any illness or disability that may affect your site actions, eg colour blindness, epilepsy, diabetes or vertigo.

You must not come to work under the influence of alcohol or drugs or take alcohol or drugs while at work.

Avoid working alone until you know the site layout. If you have to work alone ALWAYS tell your immediate supervisor:

- where you are going
- what you will be doing
- when you will be back
- that you **are** back.

Wear appropriate clothing, particularly footwear, helmet and high-visibility clothing, and tie back long hair so that it doesn't become trapped in moving parts or machinery.

 You are most at risk during your first days on site. Think – and plan – ahead.

4.1 Personal protective equipment

Personal protective equipment (PPE) does not stop accidents, but it can help to lessen their effects. Employers have a duty to eliminate the hazard and/or control the risk, so far as is reasonably practicable. PPE, therefore, represents a last line of defence for the individual.

There are many types of PPE, from helmets to footwear, each type designed to protect a different part of the body against a specific hazard. This chapter explains some of the common types of PPE.

The Personal Protective Equipment at Work Regulations require the employer to ensure that everyone issued with suitable protective clothing or equipment is told:

- how to wear and use it
- the hazards against which it affords protection
- the limitations of the protection
- how to carry out any pre-use checks
- how to maintain and store it
- how to keep it clean
- to report loss, damage or deterioration immediately it is detected (including the person to report to), and to obtain replacements
- to report any problems in its use
- when to return for replacement items with a limited life
- to sign for receipt of issue and any replacement.

In addition to the equipment intended for specific hazards, there is of course general protective clothing for work in wet and cold weather, and high-visibility clothing for roadworks.

 Thousands of head injuries sustained at work are reported each year.

Head protection

The Construction (Head Protection) Regulations 1989 require that head protection is worn when directed to do so, to comply with written site rules and at any other time that there is a risk of head injury.

If in doubt, wear your helmet.

Change your helmet:

- at the manufacturer's recommended frequency, or
- after a significant impact, or
- if it becomes deeply scratched or cracked.

Do not leave your helmet in bright sunlight – this weakens the plastic.

Foot and leg protection

Safety footwear with both toe and sole reinforcement is essential on site to prevent crush injuries to your toes and to stop sharp objects puncturing the soles of your feet.

In Britain, tens of thousands of foot, ankle and leg injuries are sustained at work each year.

4.3 Personal protective equipment

Hearing protection

Regular exposure to excessive noise causes damage to the inner ear and permanent loss of hearing. A single exposure to a very loud noise can have the same effect. Many types of ear defenders are available, from disposable earplugs to ear muffs and system helmets incorporating ear defenders. Suitable protection can be found for every situation – **WEAR IT.**

There will be occasions when you will be legally required to wear hearing protection and may be prosecuted if you fail to do so.

Eye protection

There are several types of eye protectors and it is important to wear the correct type to give the required protection.
Seek advice from your safety adviser about the eye protection you will need and only use if for the specified purpose.

Thousands of eye injuries sustained at work are reported each year.

Hand and arm protection

Gloves give protection against cuts, toxic or irritant chemicals and dermatitis, eg that caused by cement. Use barrier creams and always check you have the correct gloves to protect against the particular hazard you face. Always wash your hands before eating, drinking or smoking.

Respiratory protection

There is a wide choice of respiratory protection for dusts, gases and micro-organisms. Seek advice from your safety adviser on the appropriate type for a job. Use respiratory protection in accordance with a written method statement for a safe system of work.

Special protection

Careful selection, maintenance, certification and regular training are needed for specialist equipment including:

- compressed air escape breathing apparatus
- artificial respirators
- fall arrester and safety harnesses.

Always select these with your manager or safety adviser and ensure written safe systems of work are followed, and that the requirements of the Personal Protective Equipment at Work Regulations are implemented.

- **Always obey instructions and notices to wear PPE**
- **You should know when PPE is unserviceable and should not be used**
- **Check that you have the right type of PPE to protect against the hazard.**

People and vehicular access should be separated whenever possible

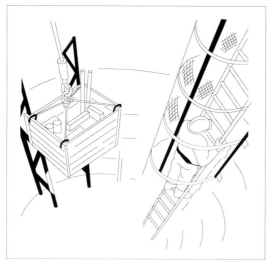

Safe access for people and their tools, equipment and materials must be provided

 A labourer fell 3 m to his death while trying to use a roof as an unauthorised access route to a scaffold.

See also Chapters 10, 16 and 22

Ground level access/egress – follow these rules:

- all visitors (including those making deliveries) should be directed to a site access control point
- access for pedestrians and vehicles should be separated wherever possible and good visibility ensured at crossing points
- pedestrian refuges must be used where provided
- adequate lighting and signs should be provided on all access routes
- access routes for vehicles and pedestrians must be kept free of obstructions.

Access/egress above ground level

Safe access to heights must be provided by use of ladders, passenger hoists or lifts that are:

- sited away from the danger of vehicle impact
- on a stable foundation
- designed, erected, inspected and maintained by competent people.

Gangways and runways should be:

- at least three boards wide if used only by people
- at least five boards wide if used for materials access
- provided with guard rails, intermediate rails and toe boards
- less than 1 in 1½ slope.

Safety lines, harnesses, cradles and bosun's chairs should only be used by trained personnel in conjunction with a written method statement for the work, which must include fixing-point details.

> **Refer to the health and safety plan for safe access arrangements on projects to which the CDM Regulations apply.**

6.1 Bottled gases

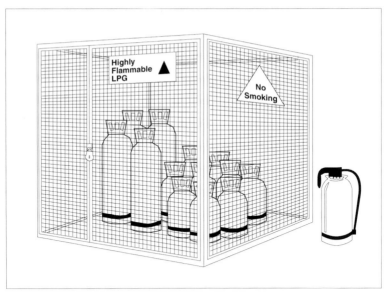

Store LPG outside in a secure well ventilated area

When LPG is used a fire extinguisher must always be available at the worksite. Staff must be instructed in emergency procedures.

Bottled gases used on construction sites include:

- liquified petroleum gas (LPG)
- oxygen
- acetylene.

All will cause fires and explosions if misused.

Always observe No Smoking signs and procedures.

 A mobile bitumen boiler overturned and set fire to two propane cylinders, which exploded, killing a 12-year-old who was playing nearby.

See also Chapter 17

You should be aware of the following safe working practices for storage, use and transport.

Storage:

- store cylinders upright in locked and secure well-ventilated labelled cages or purpose-built containers
- provide fire extinguishers nearby
- separate full and empty cylinders
- separate flammable gases from oxygen (minimum 3 m)
- site the storage cage away from buildings and excavations
- never store cylinders below ground level or in confined spaces
- never store or use an LPG cylinder in a site hut.

Use:

- ensure that cylinders are upright and cannot be easily knocked over
- check hoses, couplings and regulator for wear or damage, and ensure that flash-back arresters are used
- follow manufacturers' instructions
- always provide good ventilation
- keep bitumen boilers or similar at least 3 m from the cylinders
- return cylinders to the approved store at the end of work.

Transport:

- transport cylinders on open or well-ventilated vehicles
- secure the cylinders in an upright position
- carry an appropriate fire extinguisher
- display statutory warning notices
- ensure that drivers are trained and instructed in the hazards of carrying bottled gases.

7.1 Building maintenance

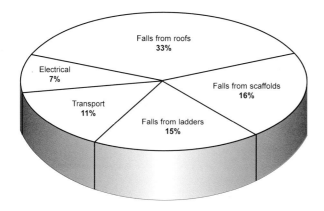

Falls from roofs
33%

Electrical
7%

Transport
11%

Falls from ladders
15%

Falls from scaffolds
16%

The five main causes of fatalities that occur during maintenance work.

Other causes **18%**

Remember: buildings may now have a health and safety file prepared for previous projects carried out in accordance with the CDM Regulations. The health and safety file must be consulted before building maintenance work takes place. The maintenance work itself may be subject to the CDM Regulations.

 More than 40 per cent of all construction fatalities occur during maintenance work.

It is essential that maintenance work is planned and executed with the same professionalism as major works. The five major causes of maintenance deaths are shown on the pie chart opposite.

Everyone involved in maintenance work must understand the hazards and the correct procedures to minimise risk. A permit to work may be required for:

- entry into confined spaces or plant and machinery
- hot work that may cause explosion or fires
- work on pipes carrying hazardous substances or dangerous gases or their residues
- mechanical or electrical work requiring isolation of power
- roofwork over a production line
- excavation work within a factory area
- work with asbestos.

Lock-off devices must be used to isolate the power supply. They must be locked off before carrying out maintenance work on or near plant and machinery etc. Permit-to-work systems must be controlled by an authorised person.

Structural stability should always be considered during maintenance and alternative means of support provided when normal supports are removed.

REMEMBER
- **plan the work and consider its impact on people nearby**
- **provide safe access, egress and working places**
- **use lock-off devices**
- **initiate permit-to-work systems**
- **it only takes an instant to be injured**
- **do not work alone unless it is necessary.**

8.1 Care with asbestos

Asbestos is a naturally occurring mineral fibre, which is normally grey in colour. Breathing in asbestos dust can cause cancer or irreversible lung damage. The more asbestos dust you breathe in, the greater the health risk.

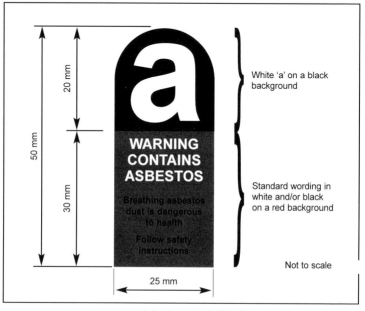

Asbestos warning label

 There is no cure for asbestos-related diseases, which kill more than 3000 people a year.

Asbestos may be found as:

- insulation to boilers and pipes
- fire protection of steelwork
- fire-protective cladding on doors, walls and ceilings
- roof sheeting
- floor tiles or ceiling tiles.

Before any work with asbestos begins, the employer must carry out a full assessment of the controls needed to protect workers and anyone else affected.

In many circumstances, asbestos work can only be carried out by a licensed asbestos contractor with notification to the HSE. A written plan for the work will be required in accordance with HSE guidelines and Approved Code of Practice for asbestos work.

If asbestos is suspected or in place, then leave it alone.

If in doubt, arrange for suspect materials to be laboratory-tested for asbestos.

The laboratory undertaking the testing of samples must be accredited to EN 45001.

Do not enter an asbestos working area unless you are:
- **trained**
- **wearing appropriate PPE**
- **instructed to do so.**

9.1 Chemicals, dust and fumes

METHANOL

Toxic by inhalation and if swallowed
Keep out of reach of children
Keep container tightly closed
Keep away from sources of ignition
No smoking
Avoid contact with skin

Highly flammable   **Toxic**

200-659-6
EEC label

Supplied by
XXXXXXXXXXXXXXXXX
XXXXXXXXXXXXX
XXXXXXXX
XXXXXXXXXXXX

Example of a supply label for a pure substance

AUNTIE MARY'S PATENT CLEANSER
contains trichloroethylene

Possible risk of irreversible effects
Do not breathe vapour
Wear suitable protective clothing and gloves

Harmful
1 litre

Mixed by
XXXXXXXXXXXXXXXXXXXXXXXXX
XXXXXXXXXXXXX
XXXXXXXX
XXXXXXXXXXXX

Example of a supply label for a manufactured preparation

See also Chapters 8 and 14

The hazards

Some substances can poison you by being directly absorbed through the skin, while others cause skin problems, eg dermatitis. Dust particles entering your lungs can scar the tissue. Some dusts cause cancer. Fumes can be fatal, especially in poorly ventilated areas. Safe systems of work must be used to limit the amount of dust and fumes produced, and to limit skin contact with hazardous chemicals.

CHIP

CHIP stands for the Chemicals (Hazard Information and Packaging for Supply) (Amendment) Regulations 2000. The objective of CHIP is to help protect people and the environment from the ill-effects of chemicals. If you know about the chemicals you use they are less likely to harm you or the environment. These regulations are amended frequently.

CHIP requires suppliers to:

- identify the hazards (or dangers) of the chemicals they supply – this is called classification
- inform their customers about these hazards
- package the chemicals safely.

Similar duties are placed on people who transport chemicals by road.

Suppliers' information includes:

- material safety data sheets
- clear labels on packaging that provide information in a standardised format.

Get hold of a copy of The Complete Idiot's Guide to CHIP, and read it.

9.3 Chemicals, dust and fumes

These symbols appear on packaging and containers to warn of the hazards associated with their contents.

Meaning

Safety precautions and medical action

Toxic/very toxic

May cause serious health risk or even death if inhaled or ingested, of if it penetrates the skin

1. Wear suitable protective clothing, gloves and eye/face protection.
2. After contact with skin, wash immediately with plenty of water.
3. In case of contact with eyes, rinse immediately with plenty of water and seek medical advice.
4. In case of accident, or if you feel unwell, seek medical advice immediately.

Corrosive

On contact may cause destruction of living tissue or burns

1. Wear suitable protective clothing, gloves and eye/face protection.
2. Take off immediately all contaminated clothing.
3. In case of contact with skin, wash immediately with plenty of water.
4. In case of contact with eyes, rinse immediately with plenty of water and seek medical advice.

Harmful

May cause limited health risk if inhaled or ingested or penetrates the skin

1. Do not breathe vapour/spray/dust.
2. Avoid contact with skin.
3. Wash thoroughly before you eat, drink or smoke.
4. In case of contact with eyes, rinse immediately with plenty of water and seek medical advice.

Irritant

May cause inflammation or irritation on immediate, repeated or prolonged contact with the skin or if inhaled

1. Do not breathe vapour/spray/dust.
2. Avoid contact with skin.
3. In case of contact with eyes, rinse immediately with plenty of water and seek medical advice.
4. In case of contact with skin, wash immediately with plenty of water.

Highly flammable

May become hot or catch fire in contact with air or is gaseous and will ignite at any ignition source

1. Protect from source of ignition.
2. Have fire precautions at hand.

The effects may be immediate or may only appear after several years.

Chemicals, dust and fumes can enter your body:

- by inhalation
- by swallowing
- through the skin.

The Control of Substances Hazardous to Health (COSHH) Regulations aim to protect workers from the effects of hazardous substances.

Hazardous substances include:

- solvents
- plaster
- fillers
- animal droppings
- concrete additives

- cement
- weedkiller
- brick dust
- micro-organisms
- ground contamination

- glues
- bitumen
- silica dust
- PCBs, eg from transformers.

COSHH requires employers to take six steps:

1. know the substances employees (including you) may be exposed to
2. assess the hazard to health they can cause
   - level of risk and degree of exposure
3. eliminate or control the hazard
   - use a non-hazardous alternative
   - limit the number of people exposed to the substance
4. inform, instruct and train employees in
   - the nature of the risk and controls to be adopted
   - reasons for using PPE
   - monitoring to be carried out
5. monitor the effectiveness of controls and initiate health surveillance where appropriate
6. keep records.

9.5 Chemicals, dust and fumes

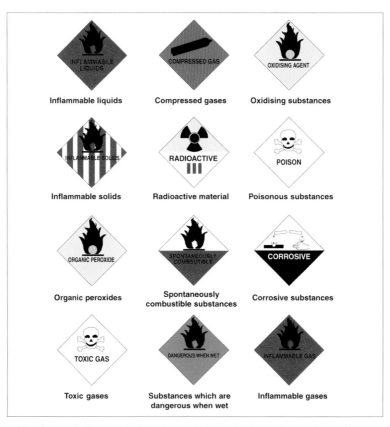

Inflammable liquids   Compressed gases   Oxidising substances

Inflammable solids   Radioactive material   Poisonous substances

Organic peroxides   Spontaneously combustible substances   Corrosive substances

Toxic gases   Substances which are dangerous when wet   Inflammable gases

The hazard diamonds illustrated above tell you the nature of the hazard of a substance. They appear on bulk tankers and on fixed installations, eg storerooms and on packaging and containers.

Remember that a COSHH assessment considers the work process, the substances involved, the risks to health and safety, who is involved and how the substance could cause harm, eg by skin contact.

See also Chapters 8 and 14

Each year the HSE produces guidance document EH40, which lists the safe working limits for fumes and dust in air. As these limits are constantly reviewed and often revised it is essential that you use the current edition of EH40.

Your duties under COSHH are to:

- take part in training programmes
- read container labels and information sheets
- follow safe working practices and COSHH assessments
- use the correct PPE
- pay attention to personal hygiene
- store chemicals and equipment safely
- report any hazard or defect to your manager
- take part in health surveillance
- know emergency procedures.

You should be aware of the risks to third parties posed by legitimate activities, or accidents, occurring within the site, for example, wind-blown paint spray or dust, or accidental pollution of rivers and drains.

Two common site hazards, lead and asbestos, are not covered by COSHH because they have their own specific regulations. The basic points outlined above still hold true, however.

The COSHH Regulations require the identification and control of ALL hazardous substances at work.

10.1 Confined spaces

Confined spaces kill

Two cases illustrate the point:

> At a reservoir site four men, all aged under 30 and physically fit, died in a surface water manhole 4 m deep. Naturally evolved carbon dioxide had displaced oxygen. No tests were made before entry. The first man down collapsed and the three other men climbed down to their deaths in futile attempts to effect a rescue.

> When an engineer collapsed in a sewer, a rescuer entered without breathing apparatus and was overcome; a second person made a similar vain attempt to reach the victims. When the rescue team from the fire brigade arrived they had to remove the two would-be rescuers before they could get to the engineer. By that time it was too late and he died.

The Confined Spaces Regulations 1997 state:

"No person shall enter into a confined space unless it is not reasonably practicable to achieve that purpose without such entry".

Furthermore: "No person shall enter a confined space unless a system has been devised assessing the risks and which makes the work safe and without risk to health". Suitable emergency measures should also be in place to enable all persons to be rescued in the event of a danger to health occurring.

A confined space is "Any place including any chamber, tank, vat, silo, pit, trench, pipe, sewer, flue, well or other similar space in which by virtue of its enclosed nature, there arises a reasonable foreseeable risk".

Typical confined spaces include:

Confined spaces are **not** necessarily small or completely enclosed.

- shafts
- tunnels
- manholes
- sewers
- box girders
- ceiling voids
- cellars and basements
- boilers and process vessels.

Even in an emergency, do not enter a confined space unless you are fully equipped to do so. If in doubt, assume the workplace is a confined space.

The three main hazards of confined spaces are:

- suffocation – lack of oxygen
- toxic atmosphere – presence or ingress of gases

eg   carbon monoxide
hydrogen sulphide
nitrogen oxides

- flammable atmosphere – presence or ingress of gases

eg   methane
petrol vapour
town gas.

In a confined space it may be difficult to lift and carry things, and movement generally is restricted and slowed down.

Do not enter a confined space until all the following conditions have been met:

- that work cannot be done without entering the confined space
- you are part of a trained team of sufficient number for the job
- you are working to and understand a written method statement, which preferably involves a permit-to-work control system
- all necessary atmosphere tests have been properly conducted and recorded, adequate ventilation is provided (or breathing apparatus is used) and continuous atmosphere tests will be undertaken
- you are equipped with adequate
  - overalls, gloves and footwear
  - breathing and head protection
  - safety harness, lighting and communications
- rescue arrangements and emergency procedures have been planned and you know them. Trained and suitably equipped person(s) must remain at the entrance of the confined space for the duration of the work.

11.1 Cranes and hoists

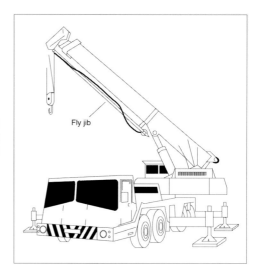

Outriggers are used to ensure crane stability

Fly jib

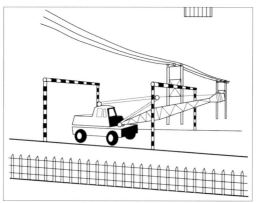

Travelling under power cables, use safety clearance goalposts

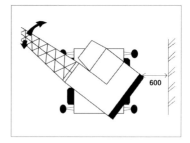

Provide at least 600mm clearance between cranes and any obstructions

600

All cranes and hoists must comply with the Lifting Operations and Lifting Equipment Regulations 1998.

All lifting operations need to be planned and managed.

Crane access

- overall height, width and swept path of crane for delivery must be checked
- approach and working areas to be as level as possible
- ground surface to be capable of taking loads
- safe height for passage underneath power lines to be determined – by reading the marker plate underneath the wire, or by phoning the electricity company – and goal posts erected.

Crane siting

- all cranes must be sited in a stable position
- maintain at least a 600 mm clearance between cranes and obstructions to prevent anyone becoming trapped
- barriers to separate cranes from overhead power lines must be at a horizontal distance of at least 6 m plus jib length from the power lines. Mark the danger area with permanent stakes or flags and high-visibility tape
- when the safe working distance cannot be maintained, contact the electricity company to investigate re-routing or disconnecting the electricity supply – this may not be simple and must be planned well in advance
- when there are several cranes on site they must be sited clear of each other to prevent possible fouling of the jibs and loads
- the working area must be kept free of unnecessary obstructions and adequate lighting provided
- the crane must be a safe distance away from excavations, slopes, underground services, soft ground etc with outriggers fully extended. Use grillages to distribute the load.

11.3 Cranes and hoists

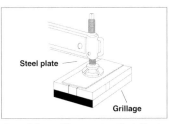

Provide grillages to distribute the force from outriggers

Protect the load and the slings by providing packs

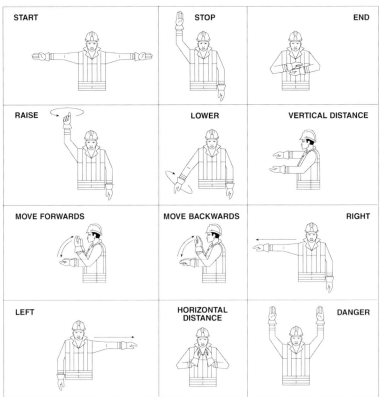

Ensure that standard lifting signals are used and that only trained banksmen are employed

Refer to the health and safety plan for safe crane working arrangements.

Crane operation

All operations must have full regard for public safety. Never lift over areas used by the public. Cranes must only be operated by trained and certificated drivers. Before starting lifting operations the driver must have documents recording the weekly inspection and the following prescribed examinations or tests:

- following any significant change or potential deterioration
- at a maximum interval of 12 months (six months if suitable for lifting persons) or as determined by the competent person
- test and thorough examination within the last four years
- inspection every six months of chains, slings and lifting gear.

These inspections and examinations must be recorded in the prescribed manner and documents must be available.

The crane must not be used if these documents are unavailable, incomplete or out-of-date.

A trained banksman (who will use only the standard signals) and trained slingers should be present. It is essential that the crane driver knows who the banksman is. An appointed person (BS 7121) must plan and monitor the lifts.

All cranes with a lifting capacity above one tonne must have an automatic safe load indicator and the weight of all loads and the lifting radii should be determined in advance of lifting. A crane must not be allowed to operate with the safe load indicator bell sounding: if it does, ensure that the crane driver stops and inform your manager.

Use tail ropes to control unwieldy loads, eg formwork or bundles of scaffold tubes. **Never use a single sling.**

Each day the crane driver should:

- inspect the whole machine including ropes, tyres and tracks, lifting gear, including chains
- check that the automatic safe load indicator and load/radius indicators are working
- put the crane through all its movements to check brake and clutch operation.

11.5 Cranes and hoists

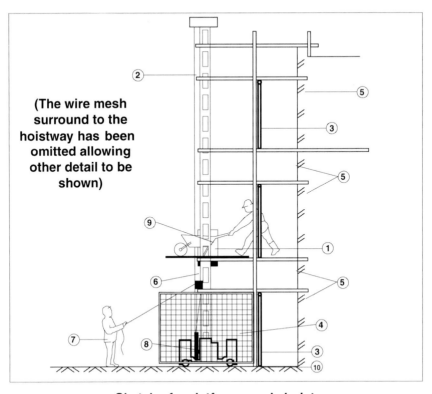

(The wire mesh surround to the hoistway has been omitted allowing other detail to be shown)

Sketch of a platform goods hoist

Safety rules for goods hoist operation

1. No passengers to be carried
2. An over-run device must be fitted
3. 2 m high gates at landings
4. The hoist must be enclosed appropriately
5. The hoist mast should be tied to the structure
6. A fall-arrester device must be fitted
7. Operator must have clear view of whole hoist
8. Fail-safe operating control must be fitted
9. Safe working load must be displayed on the platform
10. Gates to be closed before the platform is moved
11. Hoists should only be operable from one position.

Goods hoists

There are three main hazards associated with goods hoists:

- falling down the hoistway
- being struck by the moving platform
- being hit by material falling from the platform.

Hoists must be erected, extended and dismantled only by competent, trained people. Hoist operators must be aged 18 or over and be trained to operate the hoist in question. The driver must be positioned so he can see the entire hoistway.

Follow these rules for carrying materials on hoists:

- place loose materials (eg bricks) in a container or use a hoist with a cage; tall materials must be kept within the cage
- chock wheelbarrows or other mobile plant on the platform
- ensure that the safe working load is displayed on the platform and that it is not exceeded.

The following tests and records are required:

- the safe working load must be recorded in the prescribed manner
- every week the hoist must be inspected before use, and after alteration
- a thorough examination must be undertaken as detailed in the scheme of examination prepared by the competent person.

These inspections and examinations must be recorded in the prescribed manner and documents must be available.

Never allow anyone to ride on a goods hoist. There are additional requirements for passenger hoists.

12.1 Demolition

Demolition work is particularly hazardous. It should only be carried out by competent and experienced demolition contractors.

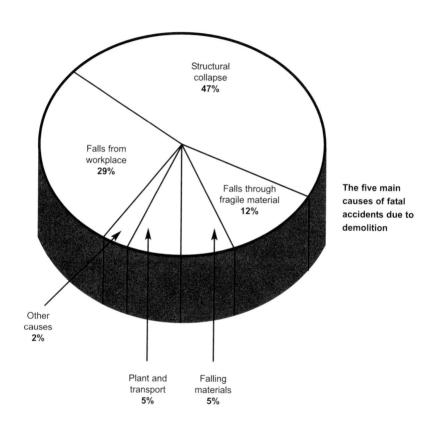

Structural collapse **47%**

Falls from workplace **29%**

Falls through fragile material **12%**

The five main causes of fatal accidents due to demolition

Other causes **2%**

Plant and transport **5%**

Falling materials **5%**

> **If in doubt about the stability of a permanent or partly demolished structure, consult a structural engineer. Always check with your manager if you are concerned.**

See also Chapters 10, 14 and 18

There are three main types of demolition:

- by mechanical means
- by explosives
- by hand.

Demolition is high-risk and should be closely supervised and carefully planned.

Do not enter demolition sites without first getting permission from both the person in charge and your manager.

Key points include:

- safe distances from the structure must be established to eliminate the hazard of debris falling on workers or the public during explosive demolition of a factory chimney or multi-storey block
- the public and workers must be protected from falling materials by the use of properly constructed boarded fans. Fans are for protection only. Do not use them for access, stacking or storage of materials
- safe places of work must be provided, complete with guard rails and toe boards. When these cannot be provided, safety harnesses must be used. Mobile access platforms are preferred
- temporary struts and guy ropes must be securely anchored and clearly marked or flagged.

 People die in demolition accidents in Britain every year.

12.3 Demolition

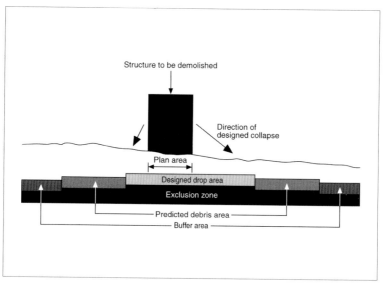

Structure to be demolished

Direction of designed collapse

Plan area

Designed drop area

Exclusion zone

Predicted debris area

Buffer area

Establishing exclusion zones when using explosives in demolition

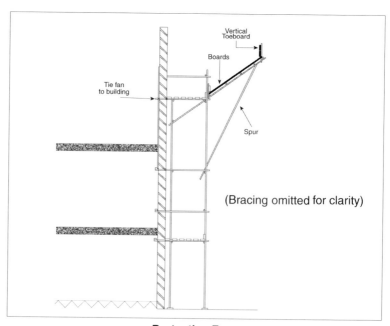

Vertical Toeboard

Boards

Tie fan to building

Spur

(Bracing omitted for clarity)

Protective Fan

Before demolition work starts:

- a written method statement must be prepared, agreed and integrated with the health and safety plan for the project
- a competent person in charge of the operation must be nominated in writing
- a competent person must inspect the structure
- particular attention must be paid to hazardous substances. If asbestos is found, it must be removed by a licensed contractor
- information about the building structure should be obtained
- the health and safety file must be consulted where one exists
- the utility companies must be contacted and all services disconnected or diverted
- existing floors that are to be used as working platforms must be suitable for that purpose
- all necessary shoring, lighting, signs etc must be planned and the work carried out at the correct time
- suitable PPE must be provided and used during operations.

During operations:

- adequate protection and safe access for the public and workers must be provided at all times, including protection from dust and noise hazards
- all the employees must be competent for this work
- all machines must be suitable, robust and placed in safe working positions on suitable ground
- overloading of existing floors (or any part of the structure) must not be permitted. If any sign of weakness that might lead to structural instability is detected refer it to your manager immediately.

13.1 Electricity

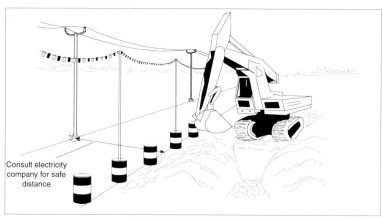

Safe working near overhead power lines requires a separation zone

There are many electrical hazards on a typical site, eg:

- 33 kV overhead power lines
- 11 kV buried or overhead cables
- power tools
- energised installations in partly completed structures.

There is no safe voltage. Even a small shock can startle you and cause you to jump back causing a slip or fall. All electrical equipment should be considered dangerous.

Voltage	Sheathing colour
25	Violet
50	White
110-130	Yellow
220-240	Blue
318-415	Red
500-750	Black

Guidance on voltages is given by the sheathing colour

 300 workers suffer major injuries from electric shock in Britain each year.

See also Chapters 11 and 26

Overhead/underground cables

Consider the location of overhead and underground cables at the planning stage of a project. Overhead wires are not insulated. Electricity can jump air gaps and current can be induced in metal structures parallel to power lines.

The precise position of buried cables must be determined from utility plans and confirmed using cable locators and hand digging. Electric cables do not lie in straight lines, they snake about within a trench. These rules should be followed:

- consult the electricity company to arrange diversions, isolation, or permit-to-work arrangements
- erect clearly marked barriers to protect overhead power lines that remain live – provide 6 m minimum clearance or as directed by the electricity company
- report any damage to the insulation of underground cables to the electricity company.

Electricity distribution and use on site:

- use portable electric tools powered by voltages of 110 V maximum, unless special prior arrangements have been agreed with your manager
- do not use home-made extension cables
- do not use multi-way adaptors or domestic 13 amp plugs
- repairs to electrical equipment must be by competent persons
- check the condition of plugs, leads, power tools and controls
- use the correct leads and sockets for the voltage supplied
- check that fuses are of the correct rating
- ensure temporary electrical systems are properly installed and tested
- plan for regular inspection and maintenance of all distribution systems, power tools and electric appliances
- use residual current devices (earth leakage circuit breakers) at the point of supply for 240 V hand-held equipment.

14.1 Environment

The Environmental Protection Act is the law. It deals with the prevention of public and private nuisance from the effects of pollution.

Protection of the environment both off and on site from pollution due to site activities must be controlled at all times.

Know your local environment! Are there hospitals, schools, day centres, old peoples' homes or residential projects adjacent to your site that may be put in danger by site activities?

The main forms of pollution emissions from sites are:

- contaminated land — disturbance, excavation, removal
- waste disposal — transport, fly-tipping, burning, bonfires
- air — fumes, gases, dusts, vapours
- water various — direct or indirect discharge of substances, effluents, spillages to drainage systems, rivers or estuaries
- noise — mechanical plant and equipment, transport, explosive blasting, piling, hammering.

The main hazards and contaminants that cause danger to human health are:

Hazard	Contaminants
Toxic by ingestion	heavy metals, phenols, coal tars, cyanide
Toxic by skin contact	oils, tars, phenols, asbestos, cement and other dusts/hazardous materials
Toxic by inhalation	hydrogen sulphide and other gases
Risk of explosion or fire	various gases and flammable or combustible materials

Prevention is better than cure.

See also Chapter 9

Main causes of contamination from site activities may occur as a result of:

- leaks and spillages from tanks, containers and pipes/hoses either during transport, storage use or disposal
- disposal of waste materials from site
- vapour or gaseous emissions from spraying or burning
- demolition of buildings containing contaminated material, eg asbestos lagging or claddings, sealants, lead in old paint, toxic waterproofing materials on or in floors, walls and ceilings
- movement and/or migration of contaminated groundwater.

What should you do?

- check if an environmental assessment has been carried out for the development. If so, familiarise yourself with the details
- read the construction health and safety plan
- check COSHH assessment for hazardous material
- check that wastes are being properly segregated, controlled and disposed of
- read *Waste Management – The Duty of Care*. A Code of Practice, available from Stationery Office.
- if in doubt, consult your site safety officer
- obtain video *Building a Cleaner Future* (CIRIA SP141V).

There are three environmental agencies operating in the UK.
England and Wales
  Environment Agency
Scotland
  Scottish Environmental Protection Agency (SEPA)
Northern Ireland
  Department of Environment (Northern Ireland)

15.1 Excavations

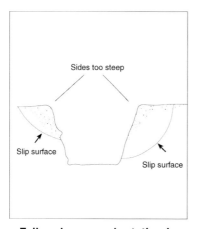

Failure by ground rotation in soft clays

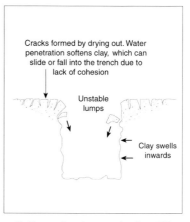

Failure due to cracks in stiff clays

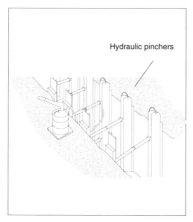

Use hydraulically operated shoring devices in preference to timber

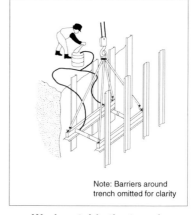

Work outside the trench whenever possible

No unsupported excavation can be considered safe, however shallow.

No ground can be considered totally stable: the ground may be inherently weak, laminated or have been disturbed previously.

Water may create instability in excavations. This can occur through:

- the action of rainwater
- changes in groundwater conditions and seepage
- erosion by water
- frost action
- drying out of soil.

It is essential that all excavations are made safe by:

- having sides battered to a safe angle of repose, or
- providing structural supports, eg trench sheets and struts, drag boxes, sheet piling or proprietary systems.

Shallow trenches may not require support if the ground is firm, provided that proper safety procedures exist and are always carried out. All trenches must either have their sides adequately supported or be battered back to a safe slope.

Where ground and support arrangements allow, install the supports before excavation to final depth. The excavation and support installation should proceed by steps until final depth is reached. Adopt safe working practices, use proprietary systems which can be installed from outside the trench or work progressively forward from existing supports.

 Eight people die in sudden excavation collapses in Britain in a typical year.

15.3 Excavations

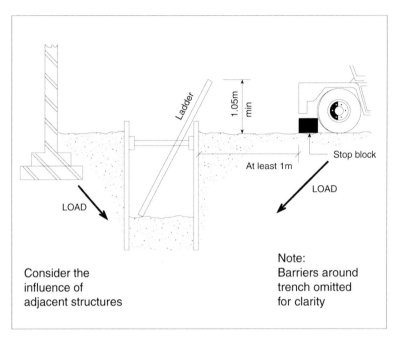

Protect the trench

Before work starts

All excavation work must be planned. Before digging begins on site, check that arrangements have been made to:

- locate underground services – check with all utility companies, look for indicator posts, valve covers and evidence of trenching, use cable and pipe detectors
- establish ground conditions and design an adequate support system
- consider the interaction of proposed excavations and existing structures:
    - will the excavation destabilise existing structures?
    - do the existing structures have adequate foundations?
    - has the ground already been disturbed?

- read the pre-tender health and safety plan
- provide edge protection for all excavations, use projecting trench sheets or other solid barriers where possible. If a person could fall more than 2 m, suitable barriers must be provided
- ensure the person directly supervising the work is fully experienced and competent in the support of excavations
- obtain the necessary drawings or sketches
- brief and instruct operatives
- provide site security, particularly in relation to preventing children getting on to the site
- establish adequate working space for plant and for spoil heaps (spoil heaps and materials should not be less than 1 m from the edge of the excavation)
- construct bridges or gangways as required
- provide sufficient ladders secured against movement
- assess need for lighting
- issue appropriate protective clothing and equipment
- protect the public (including handicapped or blind people)
- control traffic
- consider if fumes could collect in excavations.

REMEMBER
- **read existing CDM health and safety files**
- **read CDM health and safety plans**
- **update existing CDM health and safety files.**

15.5 Excavations

During the work

A competent person must inspect the excavation and its supports at the start of each shift and after significant change, eg accidental earth fall or storm, to address the following questions:

- is access to and from the workface sufficient and secure?
- are all working faces secure, wedges tight and support material free from damage?
- is there any sign of movement or deflection in the support system?
- is the soil condition as predicted? If not, what action should be taken?
- are spoil heaps an adequate distance back from the trench edge?
- are pipes, bricks and other materials, plant etc well clear of the edge so that there is no risk of them falling into the trench or of vibration causing danger to the support?
- are services that cross the trench properly supported?
- is the method statement being properly followed in installing the support? (This is particularly important in relation to the spacing of walings and struts)
- are regular tests for gases or fumes being carried out? Is ventilation required?
- has the risk of flooding been properly assessed?
- where pumping is necessary, is a proper watch being kept to ensure that fine material is not being drawn out from behind the support system?
- is resuscitation equipment available and a nominated person trained to use it?
- have all persons been instructed in evacuation procedure and the correct rescue procedure to follow if someone is overcome by gases or fumes in the trench?
- are operatives wearing safety helmets? Is any other protective equipment needed?
- is the work adequately protected and marked during the day? Is it fenced, or covered, and lit at night? Are watchmen needed?

- do gangways or bridges comply with the requirements of the Construction (Health, Safety and Welfare) Regulations in relation to width, guard rails and toe boards. Have access bridges for plant and vehicles crossing the excavation been designed by competent persons?
- are vehicle stop blocks in position?
- are vehicles/plant too close to the excavation?
- is there an agreed system of support withdrawal and have those carrying it out been properly instructed?

An inspection must also be made and recorded at least once a week.

An inspection must also be made after any fall of earth or rock, or after any other event that may have affected the excavation stability.

Alterations to planned buried service routes and the precise location of existing buried services must be recorded on "as built" drawing for inclusion in the health and safety file on projects to which the CDM Regulations apply.

A man was crushed when a trench collapsed. The trench was 1.9 m deep and was being dug by hand. No supports were available on site.

16.1 Falling

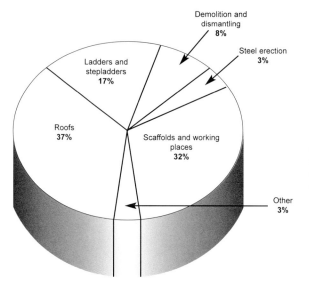

- Demolition and dismantling **8%**
- Steel erection **3%**
- Ladders and stepladders **17%**
- Roofs **37%**
- Scaffolds and working places **32%**
- Other **3%**

The workplaces from which construction workers fall to their deaths

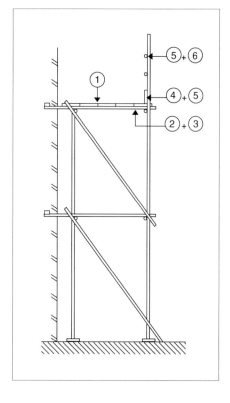

Preventing falls from an independent tied scaffold

(See checklist opposite)

See also Chapter 23

Scaffolds should be designed, erected, altered or dismantled only under the supervision of a competent person and by competent and experienced workers. Scaffolds should be inspected weekly by a competent person and the inspection recorded.

Does your scaffold adequately meet the requirements to reduce the risks of falling?

Check these points as a guide for traditional scaffolds:

1. platform three to five boards wide, depending on use
2. each scaffold board on a working platform to have at least three supports – to be determined by strength of board, but not more than 1.5 m apart
3. scaffold boards either tied down or overhanging each end support by at least 50 mm but not more than four times the thickness of the board
4. guard rails and toe boards along the outside edge and at the ends of any working platform from which people or materials could fall. Use guard rails and toe boards at inside edge to prevent people or tools/materials falling
5. toe boards at least 150 mm high, with no more than 470 mm between the top of the toe board and the guard rail or between guard rails
6. top guard rail at least 910 mm above the platform.

 In a recent five-year period there were 383 construction deaths by falling.

16.3 Falling – from roofs

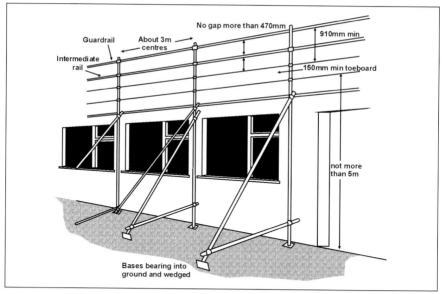

Edge protection on a sloping roof using tube and fitting scaffolding

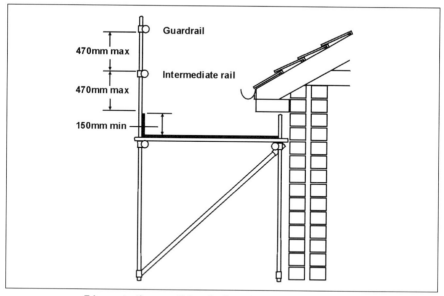

Edge protection on a flat roof using tube and fitting scaffolding

For all work on roofs, risk assessments must be prepared together with method statements and brought to the notice of those who may be at risk.

The main hazards include:

- falls from flat or sloping roofs
- falls through fragile roofs
- falls from ladders or scaffolds.

To minimise the risks, follow these rules:

- be aware of the precautions to be followed when working at heights
- display prominent permanent warning notices at the approach to any fragile roof
- never walk on fragile materials such as asbestos cement, plastics, GRP or glass. Beware – they may have been painted over or covered with insulation or dirt
- always use a planned means of access such as crawling boards or a proprietary system
- on a flat roof make sure that the edge is protected by a parapet and/or guard rails and that the roof is strong enough to support you
- where someone could fall over the edge of the roof, guard rails and toe boards must be installed or anchorage points for safety harnesses provided. Make sure that safety harnesses are worn
- prevent items falling onto people below - use brick guards, toe boards and warning notices
- keep off roofs in bad weather, eg high winds (especially if carrying sheet materials) or where there may be other hazards such as fumes or ice.

When working above ground level there is no automatically safe height, always assess the situation.

Falling 16.5

An engineer fell 4 m while descending a ladder. He was advising on the renovation of a balcony. A fully boarded out scaffold had been erected and he gained access by a ladder, which was not secured but was footed by an assistant. Unfortunately, when the engineer came to descend he did not check that the assistant still had the ladder footed securely. The assistant was caught unawares, the base slipped and the engineer fell onto the pavement.

Always follow these rules:

- secure ladders against slipping when possible by tying at the top. A second person standing at the foot to prevent slipping is effective only with ladders less than about 5 m long. For longer ladders use stakes etc
- ladders should extend approximately 1 m above the landing place or the highest rung in use, unless an alternative hand-hold is available
- arrange ways of carrying tools and materials up and down so that both hands are free to grip the ladder
- use a ladder stay or similar device to avoid placing ladders against a fragile support, eg plastic gutters
- never place ladders where there is danger from moving vehicles, overhead cranes or electricity lines
- ensure that ladders have level and firm footings – never use unsteady bases such as oil drums, boxes, planks or tower scaffolds
- do not support ladders on their rungs
- extending ladders should overlap at least three rungs
- set ladders at a slope of 4 to 1
- provide a suitable platform in ladder runs taller than 9 m
- check ladders regularly for defects – never use damaged or home-made ladders. Take damaged ladders out of use.

A demolition worker tripped over a pneumatic hose and fell 4.6m while working at an open edge on a demolition site.

A foreman fell 3 m from the edge of a floor. The guard rail had been removed so that he could use a power float. After the accident an alternative method of fixing the guard was devised so that the float could be used effectively.

A bricklayer tripped over a pile of asbestos sheets lying on his working platform. He fell 3.3 m head first over the edge to his death. Toe boards and guard rails had been removed and not reinstated.

Remember:

- keep all working places and access routes tidy and free of tripping hazards
- provide top guard rails at least 910 mm high and toe boards at least 150 mm high with no gap more than 470 mm (see page 16.3)
- provide adequate lighting
- clearly sign incomplete or dangerous scaffolds and working platforms and prevent access
- devise, implement, monitor and review safe systems of work whenever falling is a hazard
- openings should be provided with a secure cover and marked appropriately, or guard rails provided.

 Falling is the greatest single cause of death on construction sites.

17.1 Fire

The symbol shows the three conditions for fire – remove any one and the fire will stop

Flammable liquids on construction sites

Look for this symbol on container labels:

and for these warnings:

- flammable
- highly flammable
- keep away from sources of ignition
- no smoking.

 Approximately 700 people suffer serious burns on British construction sites each year.

Each year there are many fires on construction sites resulting in injuries to people and damage to property.

The Construction (Health, Safety and Welfare) Regulations require provisions for the prevention and control of emergencies, including fire. This includes:

- emergency routes and exits
- evacuation procedures
- where necessary, fire detectors, alarm systems and fire-fighting equipment.

HS G168 *Fire Safety in Construction Work* gives extremely useful guidance covering:

- how to stop fire occurring
- reducing ignition sources
- general fire precautions
- emergency procedures.

It also specifically describes what each of the duty holders under the CDM Regulations needs to do in the areas of construction activity.

The information sheet CIS51, *Construction fire safety,* is useful for construction projects with lower fire risks such as low-rise housing developments. Fire precautions and emergency arrangements should be included in the health and safety plan.

Suitable emergency routes and exits must be provided, signed and kept clear. Emergency lighting for these routes should be provided where risk assessment indicates this to be necessary.

Details of permanent emergency systems, eg sprinkler systems and fire alarms, must also be included in the health and safety file.

 Fire precautions are particularly important during refurbishment and maintenance work.

Fire extinguishers made after May 1996

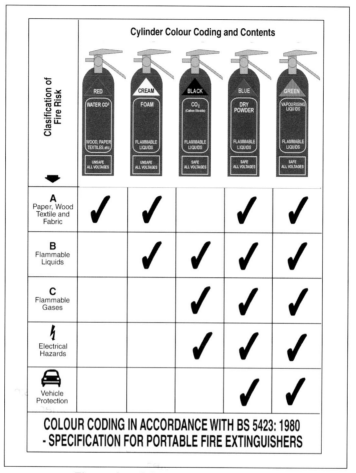

Classification of Fire Risk ↓	**Cylinder Colour Coding and Contents**				
	RED WATER CO² — WOOD, PAPER TEXTILES, etc — UNSAFE ALL VOLTAGES	**CREAM** FOAM — FLAMMABLE LIQUIDS — UNSAFE ALL VOLTAGES	**BLACK** CO_2 (Carbon Dioxide) — FLAMMABLE LIQUIDS — SAFE ALL VOLTAGES	**BLUE** DRY POWDER — FLAMMABLE LIQUIDS — SAFE ALL VOLTAGES	**GREEN** VAPOURISING LIQUIDS — FLAMMABLE LIQUIDS — SAFE ALL VOLTAGES
A Paper, Wood Textile and Fabric	✓	✓		✓	✓
B Flammable Liquids		✓	✓	✓	✓
C Flammable Gases			✓	✓	✓
Electrical Hazards			✓	✓	✓
Vehicle Protection				✓	✓

COLOUR CODING IN ACCORDANCE WITH BS 5423: 1980 - SPECIFICATION FOR PORTABLE FIRE EXTINGUISHERS

Fire extinguishers made after May 1996

Fire extinguishers made before May 1996 are entirely coloured according to the coding shown above, ie water extinguishers were red, foam extinguishers were cream, etc.

A joint code of practice, *Fire Prevention on Construction Sites,* gives detailed guidance on planning and implementing fire precautions and emergency arrangements during refurbishment on construction sites. It is available from the Building Employers Confederation.

Fire prevention is much better than fire fighting:

- flammable waste must be stored tidily prior to disposal
- do not burn rubbish on site
- flammable materials must be stored away from hazardous processes, eg welding, fabrication areas
- all flammable material stores must have warning signs
- petrol-driven plant must be switched off before refuelling and a funnel used to avoid splashes
- smoking must be prohibited within 6 m of flammable liquid and gas cylinders, NO SMOKING signs prominently displayed and good ventilation provided
- heating and cooking appliances must be properly installed
- many fires are caused by carelessness in drying wet clothes over fires/heaters.

Hot work, eg burning and welding, requires special consideration:

- consider use of hot work permits
- remove combustible material to beyond the area of sparks and spatter, or provide flameproof protection
- do not use tarpaulins as protection against sparks
- the work site must be checked after the hot work is completed, as fires can smoulder for hours.

Adequate fire-fighting equipment, fire detectors and doors must be provided.

Site personnel must know:

- the correct types of extinguisher and their location, their colour code and limitations of use
- how to use the extinguishers provided.

Each site must have a fire emergency plan. All persons must know what that plan is and the part they play in it.

18.1 Frame erection

Mobile access plant should be used whenever possible.

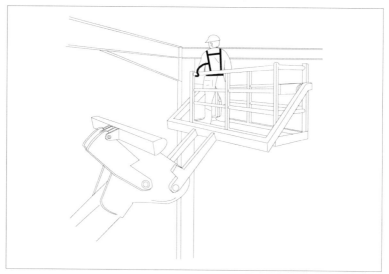

Telescopic hydraulic work platform in use for steel erection

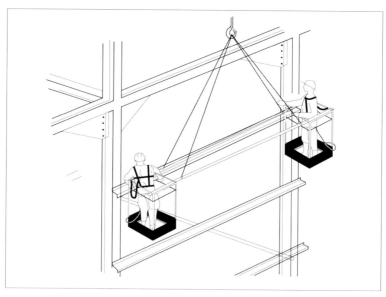

Using a sheeting rail cradle to provide safe access

See also Chapters 16 and 23

There are four main hazards associated with frame erection:

- falling
- being struck by falling tools or materials
- collapse of the partly erected structure
- adverse weather conditions.

Consideration of safety during frame erection should start at the design stage and be carried through by means of a written method statement for the site work.

Design stage – consideration should include:

- stability during all stages of erection of the structure
- the effect of the erection sequence on stability
- assessment of loading at all stages of construction, including temporary loads due to erection
- safe means of connecting components including safe access and working places – consider using remote handling where possible
- safe handling, lifting and transportation
- recognition of the practical problems of the steel erector, eg access to make connections.

The procedure for the erection of very heavy or complex members should be explained in the CDM health and safety plan. A design team member may also be required on site during steel erection to ensure there is no misunderstanding about the intended safe erection procedure.

Keep out of the area of frame erection unless it is absolutely necessary to be there. Obtain the erection supervisor's permission before you enter the area.

18.3 Frame erection

CIRIA publication SP121 *Temporary access to the workface* provides more detailed guidance on safe access procedure for frame erection including:

- mobile elevated working platforms
- telescopic boom equipment
- man-riding skips and cradles
- abseil access techniques.

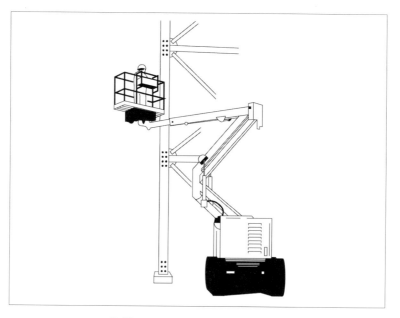

Self-propelled articulated boom

A safe system of work for frame erection should be included in the construction-phase health and safety plan.

 Wear a safety harness in telescopic boom access machines.

Erection stage – the method statement should include:

- scheme management and co-ordination, responsibilities and authority of personnel at all levels and the provision and maintenance of effective communication
- erection sequence including use of remote handling
- methods of ensuring stability at all times (including overnight) of individual components (including columns) and sub-assemblies, as well as partially erected structures
- a detailed method of erecting the structure and an erection scheme devised to ensure that lifting, initial connecting, unslinging, and final connecting are carried out safely
- procedures for work in adverse weather conditions, eg high winds and wet weather
- measures to prevent falls from height such as safe access and safe places of work. These may include special platforms and walkways, mobile towers, mobile access plant, slung, suspended or other scaffolds, secured ladders, safety harness, safety nets and supervision to ensure that all equipment is properly used
- provision of barriers such as screens, fans and nets for protection from falls of material and tools
- provision of suitable plant (including cranes) and tools and equipment of sufficient strength, capability and quantity
- contingency back-up in the event of breakdown of essential plant and equipment
- delivery, stacking, movement on site and on-site fabrication or pre-assembly
- details of site features, layout and siting of offices and stores, with notes on how these may affect proposed erection procedures.

19.1 Manual handling

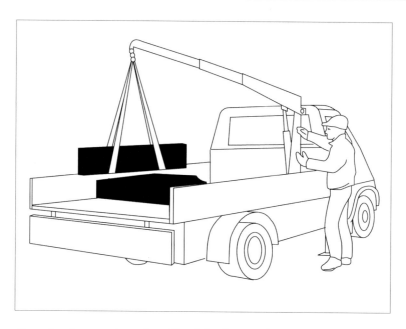

Two simple ways to reduce the risk of manual handling injuries on construction sites

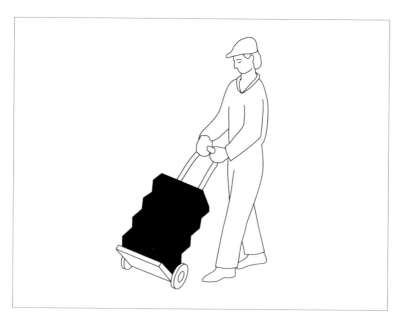

More than **25 per cent** of all reported injuries involve manual handling.

The Manual Handling Operations Regulations apply to:

- lifting
- carrying
- reaching
- pushing
- pulling
- twisting.

Employers are required to:

- avoid manual handling where reasonably practical – eg use mechanical equipment
- assess potentially hazardous manual handling that cannot be avoided
- implement measures to reduce the risk of injury as far as possible.

Designers can contribute to reductions in manual handling risks, eg by:

- designing in good access for plant, equipment and materials
- considering manual handling during future maintenance
- careful specification of materials, eg building blocks and bagged products.

Organise delivery and stacking/storage of materials to minimise lifting and carrying.

19.3 Manual handling

Bend your knees, not your back

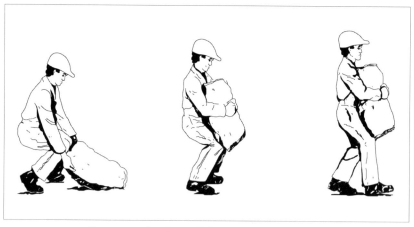

Keep your back straight, lift with your legs
and look where you are going

Keep block and bag sizes to a minimum, eg 25 kg bags of plaster.
For all awkward or heavy loads, a risk assessment must be prepared. If in doubt – ask to see it.

 Back injuries are the greatest single cause of absence from work.

See also Chapter 11

Size up the job

- are the floors sloping, slippery or greasy?
- are there obstructions or stairs?
- if the object is too heavy to lift, get help
- know where the object is to be put down.

Stand firmly

- stand close to the load
- place your feet about 0.5 m apart
- place one foot in front of the other in the direction of movement.

Bend your knees

- keep your back straight
- keep your chin well in.

Get a firm grip

- use the whole of your fingers – not just the finger tips
- keep the load close to your body
- grip boxes diagonally.

Lift with your legs

- lift by straightening your legs
- keep your back straight
- use the impetus of the lift and start moving off in the required direction.

Putting things down

- keep your back straight
- bend your knees
- don't trap your fingers
- put the load down askew and slide it into place.

Always use mechanical aids for lifting heavy and awkward objects. Get help when necessary.

20.1 Noise

Sound pressure in decibels	Situation	Sound pressure in pascals
	Peak action level, immediate irreversible damage	200
		100
140	– Jet at 30 m	
130	– Threshold of pain	
	– Pneumatic breaker (unsilenced) at 1 m	
120	– Pneumatic digger	10
	– 600 HP scraper at 2 m (pass by)	
110	– Rock drill	
	– Diesel hammer driving sheet steel at 10 m	
100	– Scabbling	1
	– 7 HP road roller on concrete at 10 m	
95	– Concrete pouring	
90	– Second action level	
	– Drilling/grinding concrete	
85	– First action level	0.1
80	– Scaffold dismantling at 10 m	
	– 8 HP diesel hoist at 10 m	
70		
	– 5 HP power float at 7 m	0.01
60		
	– Typical office	
50		
	– Living room	

Typical sound intensities

Sound pressure is measured in pascals or decibels (dB).

The decibel scale is logarithmic and it doubles with every increase of 3 dB, ie 78 dB is twice as loud as 75 dB and 81 dB is four times as loud.

The effect of noise on the hearing is cumulative. Noise exposure is normalised to an eight-hour working day over a 40-year working life.

Whenever possible, limit noise at source.

Exposure to high levels of noise over extended periods will damage your hearing. Noise-induced hearing loss is irreversible. Also refer to CIRIA PR70 How much noise do you make?

Loud noises can cause:

- permanent damage to hearing
- your hearing to become less sensitive
- permanent ringing in the ears
- breakdown of safe and effective communication.

If, with normal hearing, you have difficulty conducting a normal conversation at 1 m, the background noise level is about 90 dB; difficulty at 2 m means the noise level is about 85 dB. Move away!

The Noise at Work Regulations require employers:

- to assess noise levels
- at the **First Action Level of 85 dB(A)** to provide employees with information about the risks to hearing and provide hearing protection on request
- at the **Second Action Level of 90 dB(A)** to control noise exposure by:
  - limiting the noise at source, as the first step
  - requiring hearing protection to be worn
  - limiting the time people are exposed to noise
- at the **Peak Action Level of 200 pascals** *(equivalent to 140 dB(A))* to control noise exposure. The Peak Action Level is relevant to single loud noises and can cause instantaneous hearing damage, eg use of cartridge tools in an enclosed space.

Using environmental legislation, local authorities have the power to serve notices that specify noise levels, limit working hours for noisy operations, or ban certain types of machines if complaints are received.

If you are served with such a notice...
CONTACT YOUR MANAGER IMMEDIATELY.

> **When you are in a hearing protection area, wear your ear defenders all the time; removing ear defenders for only half an hour in eight hours reduces the overall protection afforded by 40 per cent.**

21.1 Pressure testing

General

All pressure tests must be conducted in accordance with a written method statement, which must include the risk assessment and necessary control measures. This may form part of the health and safety plan.

Before pressure testing begins, the following must be checked for damage, correct alignment, jointing integrity and compatibility:

- pipes
- valves
- fittings and flange connections
- pipeline restraints.

Pressure testing must not be carried out until all these are correct.

Pipes and fittings should be checked to ensure that they are designed to pass the pressure test plus allowable overload.

Whenever practicable, hydraulic pressure testing should be used. Air is 20 000 times more compressible than water and the sudden release of a large volume of compressed air is in effect an explosion. Use pneumatic testing only when hydraulic testing is unacceptable.

At the pressure ranges normally encountered, the amount of energy stored in compressed air or gas is 200 times that contained in water at the same pressure and volume.

Do not exceed maximum test pressure.

All pressure testing, if not carefully controlled, can be dangerous. Sudden release of uncontrolled pressure acts like an explosion and can kill and maim. Three men died in such an explosion.

Hydraulic testing

- people working in the area should be warned before hydraulic tests begin

- before a test starts, ensure that all anchorages are in position, that concrete anchorages are adequately designed for test pressure and have developed the required strength, and that the backfill between the pipe body and the trench side is well compacted

- air valves or suitable tappings should be located at appropriate high points of the main to allow the air to escape while the pipe is being filled. If the pipeline is on a level grade, it may be necessary to bleed air off at several points to ensure complete evacuation. After the air has been evacuated, all vent holes must be plugged

- where the joints of buried pipelines are to be left uncovered until testing has been completed, sufficient backfill material should be placed over the body of each pipe to prevent movement

- each section to be tested should be properly sealed off with special stop ends designed for the safe introduction and disposal of the test water and release of air

- stop ends should be secured by adequate temporary anchors. The thrust on the stop ends should be calculated and temporary anchors designed accordingly

- the section under test should be filled with clean water, taking care that all air is displaced. Where the pipeline will be used for carrying potable water, the water used for testing should be clean and disinfected.

Never use oxygen, propane, acetylene or any flammable gas to pressure-test pipelines.

22.1 Public safety and site security

Major causes of fatalities to the public arising from construction work and lapses of site security are:

- being struck by falling objects
- falling from height
- being struck by moving plant or equipment
- falling into holes and drowning or being crushed (particularly children)
- hazards present in partially completed work. Many injuries are caused by tripping over unexpected obstacles.

Children

- construction sites are a magnet to children
- wherever possible, arrange for a company safety officer to visit local schools to warn of construction site dangers
- consult HSE guidance for further precautions to be taken.

Storms

When high winds are forecast and before every site shutdown (eg Christmas) pay particular attention to:

- scaffolding, ie check ties, sheeting, bracing, boards and all connections and foundations
- stability and security of boundary fences
- storage including sheet materials and waste materials, eg empty chemical containers
- stability of lighting towers, cranes, hoardings and all temporary works.

Measures to protect public safety must be included in the health and safety plan.

See also Chapters 11, 20 and 29

General

All employers have a general duty under Section 3 of the Health and Safety at Work etc Act to take reasonably practicable measures to minimise risks to the general public.

These include:

- secure site fences wherever practicable. Display warning signs and keep access gates locked outside working hours to prevent unauthorised access
- safety zones to separate the public from construction works. This is particularly important for roadworks and multi-storey works in city centres. In such cases the highway authority must be consulted
- sites that cannot be fenced off should have high-visibility barriers, warning signs and adequate lighting and security staff as appropriate
- material must never be thrown or dropped from a height in an uncontrolled way
- securely support and guard all excavation to limit public access – provide lighting and secure fencing for these areas
- material deliveries and site activities (eg scaffold tube hoisting) to be organised to avoid lifting operations over roads and pavements open to the public. If such lifting is unavoidable, arrange diversions and/or closures
- mobile plant and equipment is to be immobilised when it is not in use. Small plant, bottled gases and chemicals are to be securely stored in their approved places
- construction operation in public areas must cause minimum disruption to pedestrians and road users.

The CDM Regulations also require a principal contractor to take reasonable steps to ensure only authorised persons are allowed on site.

 One member of the public is killed by construction activities every month on average.

23.1 Scaffolding – mobile

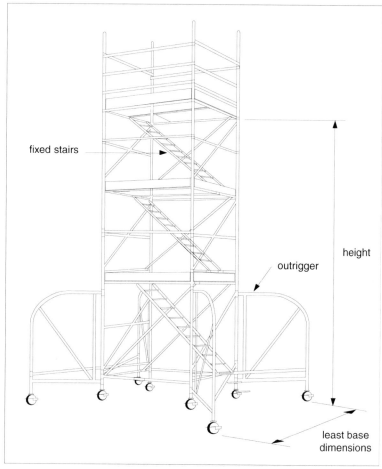

A typical tower scaffold

There are two main hazards associated with both mobile and fixed scaffolding:

- people and/or materials falling from the scaffold
- overturning or collapse of the scaffold.

Chapter 16 deals with falling. This chapter deals with precautions against scaffold collapse or overturning.

See also Chapter 16 and 27

Tower scaffolds

- the scaffold is to be erected and used in accordance with the manufacturer's instructions and the safe working load restrictions are to be observed
- the tower must rest on a firm base on level ground. Use the outriggers and ensure the castors are locked before the tower is used
- the ratio of height to least base dimension should not exceed 3:1 for outside work, or 3.5:1 for inside work, unless tied to suitable fixed points
- never use a ladder on or against the top platform of a tower scaffold. Use only internal ladders/stairs for access to the working platform – never use the scaffold framework as a ladder unless it is purpose-designed
- tower scaffolds must not be used outside during winds of   force 4 or greater unless securely tied
- the tower is to be securely tied to the structure whenever it is to be used for grit-blasting, heavy drilling, or it is to be sheeted out
- overhead power lines and obstructions must be considered when constructing, using or moving the tower
- the tower must never be moved with people or materials on the working platform
- the tower must only be moved from the base
- once the move is completed the outriggers must be re-engaged and the castors locked
- should not be used unless they have been inspected by a competent person at the start of each shift.

 One person dies in a scaffolding accident every fortnight on average on a British construction site.

23.3 Scaffolding – fixed

This section applies to conventional tube and fitting scaffolds.

Support the standards on a secure base

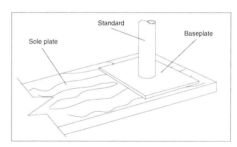

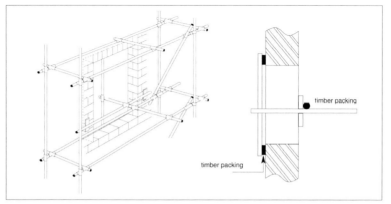

A through tie

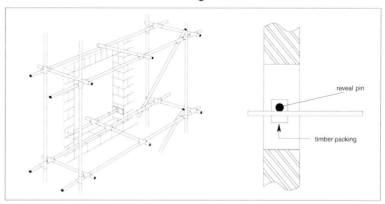

A reveal tie

Removal, renewal or alteration of ties should be carried out only by a competent scaffolder.

General access scaffold

- for scaffolding to be safe, it must be stable and designed to take the anticipated load. This depends on a secure base, a suitable height in relation to its mean base dimension or being tied to a permanent structure to prevent movement
- erected scaffolds should conform with BS 5973 and the Construction (Health, Safety and Welfare) Regulations.

Base

- level wooden sole plate (positioned from outside to inside standard) surmounted by steel base plate with central spigot to prevent displacement of the standard
- daily check to ensure no undermining, slippage or settlement.

Standards (upright tubes)

- must be vertical with staggered joints made by internal expanded fittings.

Ledgers (running longitudinally parallel to the structure)

- joined by externally placed clamp (sleeve coupler) staggered to ensure joints do not fall in the same bay
- ledger secured to standards with right-angle couplers (load-bearing).

Transoms (tubes placed on ledger at right angles to the structure)

- these complete the square of scaffolding and are used as supports for boards
- there must be a transom provided adjacent to every standard joined with right-angle load-bearing couplers
- intermediate transoms provided to support boards may be joined with wrap-over fittings (non-load-bearing).

23.5 Scaffolding – fixed

Ledger brace

- a tube running diagonally from upper front to lower rear ledger in alternate bays of scaffolding
- load-bearing couplers must be used
- may run from standard to standard (use load-bearing swivel coupler)
- may all run in same direction or in reverse of each other (dog-leg bracing).

Longitudinal facade or face brace

Fitted from base to top of scaffold at a diagonal across the face to prevent bowing of the face or sway movement.

Ties

Ties must be provided to secure the scaffold to the structure.
Only 50 per cent of ties in a scaffold may be reveal ties.

- through tie – a tie assembly through a window or opening in a wall
- reveal tie – tube placed into window reveal with screw fitting
- box tie – made by encompassing column or outer part of permanent structure
- drilled anchorages – female part fitted to suitably sound part of the structure, male part attached to scaffold to tie to the building
- where ties do not prevent inward movement this must be done by abutting the transoms to the face of the building
- ties must not be fitted to external decorative fittings, eg downpipes or balustrades.

Working platforms

Must be even and fully boarded – width depends on usage of scaffold.

Toe boards and guard rails

Required on all working platforms, access ways, stairways and landings, where a fall of more than 2 m is possible.

Access to working platforms

Access normally by ladders or stairways. Ladder must be tied and at an angle of 4:1.

Loading

Loading should be vertically above the standards where possible or at specially strengthened loading bays.

NOTE: The wall thickness of scaffold tubes can vary – eg UK tube is thicker than European tube – so load-bearing capacities can vary. Tubes of different thicknesses or of different metals (eg steel and alloy) should not be mixed in a designed structure.

REMEMBER

No scaffold – fixed or mobile – should be used unless it has been inspected by a competent person:

- at the start of each shift and a report prepared within the previous seven days
- after exposure to adverse weather conditions that may have affected its strength or stability
- after any substantial modification or alteration.

All inspections must record the prescribed information.

24.1 Site investigation

Initial survey

Before going to site consider the hazards you may encounter. These can include:

- surveying on live roads
- contaminated land
- dilapidated buildings
- confined spaces
- old cellars, walls or shafts
- unstable ground or pits
- live services
- asbestos.

Consider which hazards you may meet and how they could affect you, check the health and safety plan and implement it.

Site investigation

Drilling and digging trial pits may expose the ground investigation team to hazards including:

- contaminated land – COSHH assessments required
- collapse of ground
- methane (or other gas) pockets – the agreed safe system of work must include gas detection, testing and emergency procedures
- underground services.

The following general rules apply to site investigation:

- there should be a minimum of two drilling crew per rig
- never enter an unsupported trial pit
- appropriate lighting, barriers and warning signs must be provided at the end of each working day
- staff must wash their hands before eating or drinking
- eating, drinking and smoking should only be allowed in designated clean areas, eg site welfare facilities
- adequate storage must be provided for dirty or wet clothing and PPE.

Existing services include:

- overhead and underground electric and telephone cables
- water
- sewerage.
- gas
- chemical pipelines
- cable TV

Standard colour codes are gradually being adopted for services. Ask your manager.

All services present hazards. The first essential move towards avoiding danger from overhead or underground services is to contact the local offices of the relevant company – British Telecom, British Gas, electricity, water and cable companies and highway authorities. Obtain as much information as possible about the location of cables, lines and pipes. Thereafter, it is important to maintain close liaison with the companies as long as the work is in progress.

Erect goalposts and barriers to protect overhead services and use cable locators to pinpoint the location of underground services. After use of locators (by trained personnel), trial holes should be carefully dug, using hand tools, to confirm the position of buried services.

Once underground services have been located it is important to identify them correctly. Consult the service companies for confirmation and to record their location, type and depth permanently. Ensure the health and safety file is updated with these records.

The use of hand-held power tools and mechanical excavators too close to underground services is a major cause of accidents.

25.1 Site set-up

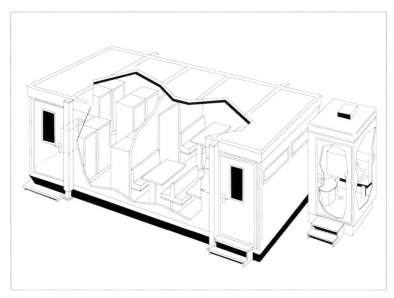

A basic welfare facility for sites

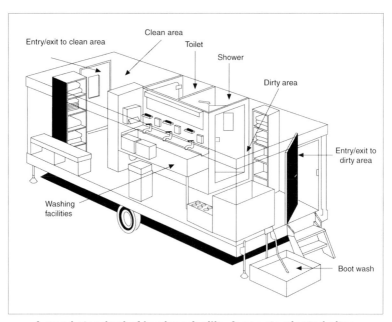

A good standard of hygiene facility for contaminated sites

See also Chapter 5, 22 and 23

Site organisation

Health, safety and welfare standards are reinforced by the early provision of:

- adequate canteen, hygiene and first-aid facilities and toilets
- a secure perimeter fencing to the site
- clearly defined site access and vehicle routes
- vehicle washing facilities
- all necessary PPE and adequate storage space for PPE
- clearly defined storage, office and working areas
- emergency procedures and statutory notices posted in highly visible places
- a mobile phone until fixed phones are installed.

Notify the emergency services of site location, activities and access.

Major sites should have a fully equipped first-aid room, but every site must have:

- soap, towels and clean hot water for washing
- at least one first-aid box
- small portable first-aid kits for people who do not have easy access to the first-aid box.

First-aid boxes should have a list of contents inside the lid.
Your company must provide an adequate number of fully trained first-aiders or appointed persons, who must be readily identifiable.

All employees should receive site induction training including details of the site layout, site rules and emergency procedures. Visitors should always be accompanied. Do not enter a site alone without first informing others.

25.3 Site set-up

The Construction (Health, Safety and Welfare) Regulations

These Regulations contain the standards for welfare facilities on construction sites. Schedule 6 of the Regulations sets out the welfare facilities required including:

- separate toilets for men and women
- clean hot and cold water for washing, with soap and towels (or other means of drying hands)
- toilets and washing facilities should be adequately ventilated and lit
- drinking water and cups
- clothes-drying facilities
- rest areas including provision for non-smokers
- means for boiling water
- arrangements for preparing meals.

26.1 Small plant and equipment

The correct PPE must be used with small plant

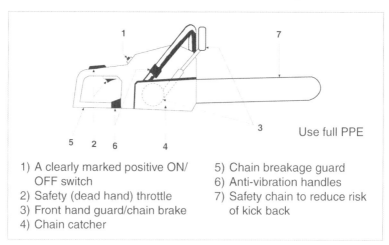

Use full PPE

1) A clearly marked positive ON/ OFF switch
2) Safety (dead hand) throttle
3) Front hand guard/chain brake
4) Chain catcher
5) Chain breakage guard
6) Anti-vibration handles
7) Safety chain to reduce risk of kick back

Chain saw guards

 20 per cent of all construction fatalities occur as a result of accidents with mobile plant and equipment.

See also Chapter 4

Small plant and equipment 26.2

All small plant and equipment is potentially dangerous.

Small plant and equipment frequently involved in accidents includes:

- saws
- abrasive wheels
- cartridge-operated fixing tools
- compressed-air tools.

Cutting, drilling, grinding, punching or sawing with small plant is dangerous.

Small plant and equipment:

- affects others working nearby
- must be used only by competent personnel
- must be kept clean and well maintained
- must have all guards fitted and effective
- can create noise and dust
- requires the wearing of appropriate PPE
- should be kept in locked stores when not in use. Stores must be clean, dry, well lit and well ventilated, and have an issue/recovery control system
- fuels should be kept and transported in suitable, properly labelled containers.

Maintenance

All plant and equipment requires regular maintenance.
Daily checks are part of these procedures and should be monitored to ensure they are done.

Examples of daily checks on chainsaws

- stop switch – works
- guide bar and sprockets – not broken
- chain brake – works
- lubrication system – full.

26.3 Small plant and equipment

All plant and equipment must meet the requirements of the Provision and Use of Work Equipment Regulations. These regulations place duties on employers to:

- ensure that equipment is suitable for its intended use
- consider the working environment when choosing equipment, eg waterproof, robust equipment for outdoor use, or intrinsically safe electrical equipment for use in sewers
- maintain equipment in good order
- train and inform staff in the use of work equipment, including how to perform daily checks and report defects.

There are also specific duties placed upon the employer to:

- guard dangerous machinery parts, eg circular saw blades
- provide adequate lighting in which to use the equipment
- ensure controls are working and easy to use, and that there are no 'home-made' modifications
- provide a means by which equipment can be isolated from all power sources for maintenance or adjustment
- protect workers from machinery parts at high or very low temperatures, eg ground-freezing plant.

Where training is required before operating tools or equipment, it should normally be certificated so that an employee can prove that they have been trained. Examples of equipment requiring certificated training include:

- cartridge tools
- chainsaws
- abrasive wheels
- burning equipment.

REMEMBER
Abrasive wheels must be changed only by those who have been trained and certificated to do so.

27.1 Temporary works

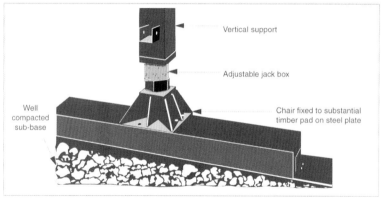

Vertical support

Adjustable jack box

Well compacted sub-base

Chair fixed to substantial timber pad on steel plate

A good foundation detail for a load bearing base

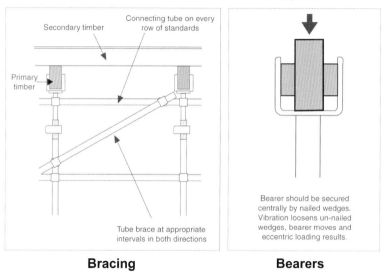

Secondary timber

Connecting tube on every row of standards

Primary timber

Tube brace at appropriate intervals in both directions

Bracing

Bearer should be secured centrally by nailed wedges. Vibration loosens un-nailed wedges, bearer moves and eccentric loading results.

Bearers

A temporary works co-ordinator should be appointed before work begins. He will be responsible for checking and signing for each stage of the works.

Temporary works include:

- falsework and formwork
- cofferdams and their bracing systems
- temporary supports to unstable structures

- temporary bridging
- scaffolding
- support to excavations
- temporary electrical supplies.

Design

- all temporary works must be designed by competent people
- calculations, drawings and sketches to explain and illustrate the temporary works design must be produced
- an independent design check should be carried out also considering any effects on the permanent works.

Method statement

A written method statement for construction and use of the temporary works is essential. It must include:

- the name of the person in charge
- detail of the design and loading limitations
- construction details and erection and loading sequence
- a specification for plant, materials and methods to be used
- details of supervision, inspection and checks to be made
- design tolerances, eg deflection/elongation
- any formal permits to load or dismantle that are required
- striking times and sequence
- dismantling sequence.

The method statement must be checked and signed, and incorporated into the construction phase health and safety plan.

There must be no unauthorised departure from temporary works design or method statements. Authority must only be given in writing by temporary works designer (co-ordinator).

28.1 Vehicles and site transport

Never hitch a lift on mobile plant

Avoid working close to plant and equipment whenever possible

 Every year many people are killed or injured on site by reversing vehicles.

See also Chapters 5 and 113

Large mobile plant

- must use planned site entry and exit points only and obey appropriate traffic control procedures
- should use separate routes from pedestrians where practicable
- should be operated only by trained, authorised and licensed drivers aged 18 years or older
- must comply fully with the Road Traffic Act when it travels on public roads (tax, number plates, lights, brakes etc)
- should have an instruction book detailing the driver's daily checks and routine maintenance
- should have amber rotating lights and reversing siren
- reversing areas should be controlled by a banksman and non-essential personnel excluded
- when tipping into or running alongside excavations must be provided with stop blocks and scotches
- must not be overloaded
- carry only well-secured loads
- must observe site speed limits – these must be clearly signed
- must be immobilised when not in use
- should be parked on level ground, in neutral, with the parking brake applied
- may deposit mud and debris on public roads. Appropriate road cleaning arrangements must be provided
- has limitations on the gradients and cross-slopes on which it can safely operate. These must be observed.

All plant and powered equipment must be operated only by the trained people authorised to use it.

29.1 Working on live roads

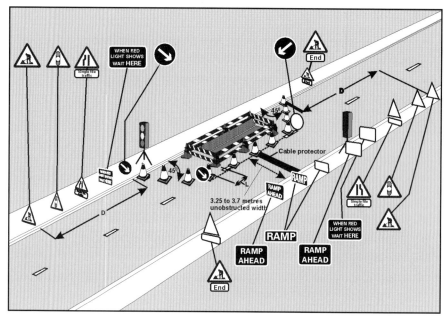

Traffic control on a single carriageway, using portable traffic signals

Traffic safety measures at roadworks are covered by
Chapter 8, Volumes 1 and 2 of the *Traffic Signs Manual* and
in a simpler form, by *Safety at Street Works and Road
Works*. Both are available from The Stationery Office.

See also Chapter 28

Planning stage

- local authorities and police must be consulted – and their requirements regarding the timing of the works implemented
- consult *Traffic Signs Manual* for details of signing requirements and size of safety zones
- decide how many of the following will be required:

- traffic signs   – road lamps
- cones   – information boards
- barriers   – site lighting.

- consider traffic management and control systems
- access must be planned to eliminate dangerous movements of site traffic (eg reversing of vehicles) and personnel (eg crossing dual carriageways)
- will the work be completed in good visibility? If not, adequate lighting must be provided
- all workers must wear high-visibility clothing
- vehicles should be equipped with amber flashing beacons
- minimum lane sizes and provision of adequate safety zones may result in the need for road closures.

On site

- define the working area in the live road/footway
- define the working space – this includes the areas for storage of tools and equipment and space to move around
- provide a safety zone – an area to separate the work from the traffic – keep it clear of all work, materials storage and people and clear of the working radius of all plant
- work must be undertaken by certificated supervisors and certificated operatives in accordance with the New Roads and Street Works Act.

30.1 Working over water

Combined harness and life jacket   **Rescue line**

Where there is a risk of people falling into water, provision must be made for:

- signs to warn of deep water
- edge protection, including guard rails and toe boards
- suitable rescue equipment, and training and instruction in its use
- keeping platforms and ladders clean and clear of debris, slime and tripping hazards
- safety nets to arrest falls where standard working platforms or harnesses cannot be provided
- safety lines and harnesses to be used in conjunction with temporary ladder access over water
- site lighting.

The wearing of some types of protective clothing can increase the risk of drowning. Non-slip work boots are preferable to wellington boots; chin straps should not be worn on safety helmets.

 Boats used for transporting people to/from their place of work must be of suitable construction and size, properly maintained and controlled by competent people.

Rescue equipment

- lifebuoys with rescue lines should be provided at intervals along the site and positioned conveniently for use in an emergency. Floating grab lines should be attached at   intervals to structures or floating plant
- personnel must wear lifejackets, or buoyancy aid equipped with a whistle and (during darkness) lights
- before the start of each shift all rescue equipment must be checked by a competent person to ensure it is present and in good condition
- enough personnel should be present who are trained in the use of rescue equipment and emergency procedures.

Rescue boats

- on tidal water or fast-flowing rivers, a power-driven boat should be provided and properly equipped, including lifebuoy with buoyant rope. In certain circumstances, eg some canals and inland waterways, a rowing boat may be sufficient
- where rescue boats are required during the hours of darkness, swivelling searchlights should be fitted
- must be manned continuously during the period that any persons are working over water and when on standby
- must be manned by competent boatmen who are also trained first-aiders
- should not be used for other purposes.

 75 per cent of drownings occur in relatively quiet waters like ponds, reservoirs and rivers.

31.1 Working alone

Never work alone in the following situations:

- in confined spaces
- over or near water
- on live electrical equipment
- in derelict or dangerous buildings.

If possible, avoid working alone in the following situations:

- on live roads
- in empty buildings
- on roofs
- near demolition work.

Although there is no general legal prohibition on working alone, the hazards that all workers face are increased when there is no one else to give a warning. Also, because there is no one else to help in the event of an accident, the risk of serious injury becomes more likely from:

- tripping, slipping and falling
- becoming trapped or crushed
- electrocution
- asphyxiation.

Plan your work

- risk assessments must be carried out for all lone working where hazards might exist
- evaluate the work to be done – are the workplace and access to it safe? Is more than one person required?
- will any lifting be necessary – can one person do this safely?

Think carefully before you work alone.

See also Chapter 33

Communicate

- leave details of your movements at a designated place – state where you are going, when you will be there, when you expect to finish and where you will go next
- report in by phone or radio at regular pre-set intervals
- always report in when you leave a site – this is particularly important when you are not returning to your office or normal site base
- if there are other activities in the area make sure your presence is known to those in charge
- take a mobile phone with you whenever possible.

Alarms

Alarms are available that emit a loud emergency siren if the wearer does not move regularly. They are therefore triggered if the wearer is knocked unconscious. They are only effective if someone is within hearing range of the alarm. You should also consider carrying a personal attack alarm.

First aid

- take a small first-aid kit with you
- if you habitually work alone, attend a first-aid course.

Assault

- never admit a stranger into a building if you are working alone
- remember the primary aim is to get away from a dangerous situation
- do not be aggressive – avoid confrontation
- never think that it could not happen to you.

32.1 Dealing with hazards

As soon as you arrive on site, find out:

- about the site safety organisation and who the site safety supervisors are
- exactly who you report to, and who and what you are responsible for
- how to report hazards that are beyond your control
- where to get further advice
- what personal protective equipment you are required to use, when and where and how to carry out any pre-use checks
- where the first-aid room is and who the first-aiders are
- fire precaution and emergency procedures and services
- bomb warning contingency plans.

Think ahead. Anticipate the safety issues you will face and know the contingency plans, including emergency plans.

Find and read the risk assessments that your employer has prepared for the work.

Read the site notices and act upon them when necessary.

On projects where the CDM Regulations are applicable, find out:

- where the construction-phase health and safety plan is kept and what it contains
- how you will be made aware of changes in the health and safety plan
- how you can contribute to the development of the health and safety plan.

If in doubt – find out!

From time to time you will encounter unsafe conditions or unsafe actions. What can you do about them?

Eliminate or reduce hazards, for example, by:

- reporting unsafe practices
- reporting unsafe workplaces
- suggesting alternative means of safer construction
- taking action to stop unsafe activities until the situation has been made safe.

Eliminate or reduce risks, for example by:

- wearing appropriate protective clothing and safety equipment
- obeying warning signs and notices
- not putting yourself at risk, even if others are in danger.

List contacts and telephone numbers for use in emergency.

When facing a dangerous situation you MUST act immediately to prevent injury to anyone in the vicinity.

33.1 Dealing with accidents

EMERGENCY AID

1 RECOGNISE A LACK OF OXYGEN

Arising from
ELECTRIC SHOCK
DROWNING
POISONING
HEAD INJURY
GASSING etc

May be causing
UNCONSCIOUSNESS

NOISY OR
NO BREATHING

ABNORMAL COLOUR

2 ACT AT ONCE

SWITCH OFF ELECTRICITY, GAS etc
ONLY REMOVE CASUALTY TO PREVENT
FURTHER INJURY
SEND SOMEBODY FOR HELP

GET A CLEAR AIRWAY ...
REMOVE ANY OBSTRUCTION ... then

TILT
HEAD BACK

LIFT
JAW

BREATHING MAY RESTART ... IF NOT ...

3 APPLY RESCUE BREATHING

START WITH FOUR
QUICK DEEP BREATHS

SEAL NOSE AND
BLOW INTO MOUTH

or

SEAL MOUTH AND
BLOW INTO NOSE

KEEP FINGERS ON JAW
BUT CLEAR OF THROAT

MAINTAIN HEAD
POSITION

AFTER BLOWING INTO
MOUTH or NOSE
WATCH CASUALTY'S
CHEST FALL AS
YOU BREATHE IN

REPEAT EVERY 5 SECS

**AFTER FIRST FOUR
BREATHS TEST FOR
RECOVERY SIGNS**
1. PULSE PRESENT?
2. PUPILS LESS LARGE?
3. COLOUR IMPROVED? **PULSE POINTS**

4 IF NONE, COMBINE RESCUE BREATHING & HEART COMPRESSION

PLACE CASUALTY
ON A FIRM SURFACE

COMMENCE
HEART COMPRESSION

HEEL OF HAND ONLY
ON LOWER HALF OF
BREASTBONE,
OTHER HAND ON TOP
(FINGERS OFF CHEST)

BREASTBONE

HEART

KEEP ARMS STRAIGHT
AND ROCK FORWARD
TO DEPRESS CHEST
40mm

APPLY 15
COMPRESSIONS ONE
PER SECOND ... then
GIVE TWO BREATHS

RE-CHECK PULSE ...
IF STILL ABSENT
CONTINUE WITH
15 COMPRESSIONS
TO TWO BREA

IF PULSE RETURNS
CEASE COMPRESSIONS
BUT CONTINUE
RESCUE BREATHING

Rescuers should not put themselves in danger. Another casualty often reduces the chance of providing rapid assistance to the first injured person.

Should an accident occur, follow these rules:

- send for a first-aider and/or doctor or ambulance
- separate the cause and the victim if possible, eg switch off electricity supplies, turn off powered plant before assisting the casualty
- move a casualty only to prevent further injury
- check the heart and breathing; give emergency aid as necessary
- stop any bleeding; raise the injured part and apply pressure
- keep the victim warm and reassured
- take care not to become a casualty yourself
- do not remove or disturb evidence
- tell your manager
- accidents to self-employed people must be reported to the person in control of the workplace
- specific types of accidents must be reported to HSE.

Reporting accidents to HSE

From April 1, 2001, there is a new procedure for reporting accidents to HSE.
By post to: Incident Contact Centre, Caerphilly Business Park, Caerphilly CF83 3GG.
By telephone: 0845 300 9923
By fax: 0845 300 9924
By e-mail: riddor@natbrit.com
or: to your local HSE office within 10 days using form F2508 or F2508A

First aid

First-aiders must be readily identifiable – preferably by wearing a distinctive safety helmet. Ensure you know the first-aiders on your site.

You should consider taking a first-aid course. Not only would you learn how to relieve the suffering of an injured colleague, it would help you to stay calm and in control when an accident happens on your site.

Find out whose job it is to report accidents to HSE on your site. It could be yours!

34.1 Accident investigation

Note all the facts immediately and add these records to your detailed site diary.

If you were a witness or had any responsibility for the work being undertaken when the accident took place you will be required to submit a written report.

Accident investigation has many benefits, including:

- identifying the underlying basic causes
- preventing recurrence of similar accidents
- identifying training needs
- providing information in case of litigation.

An accident will usually be investigated by:

- the senior site manager
- the company safety adviser
- an HSE inspector may be involved.

You may be asked to help their investigation by providing factual information and possibly your opinion.

HSE inspectors have powers to require the accident location to remain untouched until their investigations are complete.

An investigation by an HSE inspector may lead to criminal proceedings. You may be required to give evidence and your report may be used by the Inspectorate and by your company to help establish the cause of the accident. Remember it is an offence to lie or to cover up the facts.

To help you prepare your report, if you witness an accident or are nearby when an accident occurs or had any element of managerial control over the work:

- take immediate notes of what you saw and heard
- make sketches of the accident location
- if appropriate, take photographs of the scene and any relevant detail, eg a broken piece of equipment or a missing guard rail. Consider other people's feelings and explain what you are doing
- identify witnesses who actually saw the accident happen – not just those who were present when the accident occurred
- separate fact from hearsay and opinion
- identify, inspect and put aside tools, equipment and materials being used by the injured person(s) at the time of the accident
- check if the work involved a written method statement (other than for rescue) and whether it was being followed
- after your investigation, leave the scene as undisturbed as possible.

Your report will be added to your firm's accident/injury records and provide useful information for future training programmes, accident prevention measures, and for improving company policy and working arrangements.

35.1 Bibliography

Source materials

These materials were used during preparation of the three editions of this handbook. Not all are publicly available.

- Balfour Beatty Construction Ltd, *Health and Safety Update*, book revised 1997; CD ROM revised 1998.
- Building Employers Confederation, *Construction Safety*, 1999.
- Construction Industry Training Board, GE 706, *Site Safety Simplified*, 1998.
- Health and Safety Executive, *Health and safety explained*, 1989.
- Health and Safety Executive, *Safer working in tunnelling*, 1989.
- Health and Safety Executive, HSG33, *Health and safety in roof work*, 1998 (ISBN 0 7176 1425 5).
- Ove Arup Partnership, *Arup Site Safety Handbook*, 1999.
- Royal Institution of Chartered Surveyors, *Surveying Safely*, 1991.
- Stationery Office, *Safety at Street Works and Road Works. A Code of Practice*, 2001 (ISBN 0 11 551958 0).
- Symmonds, *Group Health and Safety Manual*, 1997.
- Thomas Telford, *Construction Safety Handbook*, 2nd edn, 1996 (ISBN 0 7277 2519 X)
- WS Atkins, *Safety Handbook* and *Health and Safety Update*, 1998.

This bibliography includes publications that the author referred to in preparing the handbook in addition to the source materials listed. Readers should be aware that some of these publications are updated from time to time and should check that they are using the most recent edition. Guidance that at the time of publication is being revised, is unavailable or has been withdrawn is shown by a bullet point ▫ and not •.

General

- CITB, GE 700, *Construction Site Safety Course, Safety Notes*, 1996.
- HMSO, *The Health and Safety at Work, etc. Act 1974*, 1974 (ISBN 0 10 543774 3)
- HSC, *Annual Report 1996/97*.
- HSC, L24, *Workplace health, safety and welfare. Workplace (Health, Safety and Welfare) Regulations 1992. Approved code of practice and guidance*, 1996 (ISBN 0 7176 0413 6).
- HSE, HSG65, *Successful health and safety management*, 1997 (ISBN 0 7176 1276 7).
- HSE, HSG150, *Health and safety in construction*, 1996 (ISBN 0 7176 1143 4).
- RoSPA, 1S 13, *Construction Regulations Handbook*, 13th edn. Stationery Office, SI 1999/3242, *The Management of Health and Safety at Work Regulations 1999*, 1999 (ISBN 0 11 085625 2)

Chapter 1 Your responsibilities

- HMSO, *The Health and Safety at Work, etc. Act 1974*, 1974 (ISBN 0 10 543774 3)
- HSE, CIS1, *General legal requirements*, 1988.
- HSE, INDG103, *It's your job to manage safety*, 1991.
- Stationery Office, SI 1999/3242, *The Management of Health and Safety at Work Regulations 1999*, 1999 (ISBN 0 11 085625 2)
- Thomas Telford, *Construction Safety Handbook*, 2nd edn, 1996 (ISBN 0 7277 2519 X)

Chapter 2 Construction-related regulations

- CIRIA, R145, *CDM Regulations – case study guidance for designers: an interim report*, 1995 (ISBN 0 86017 421 2).
- HMSO, SI 1996/1592, *The Construction (Health, Safety and Welfare) Regulations 1996*, 1996 (ISBN 0 11 035904 6).
- HSC, *A guide to managing health and safety in construction*, 1995 (ISBN 0 7176 0755 0).
- HSC, *Designing for health and safety in construction. A guide for designers on The Construction (Design and Management) Regulations 1994*, 1995 (ISBN 0 7176 0807 7).

35.3 Bibliography

- HSC, L54, *Managing construction for health and safety. The Construction (Design and Management) Regulations 1994. Approved code of practice*, 1995 (ISBN 0 7176 0792 5) – under revision at the time of publication.

- HSE, CIS39, *The Construction (Design and Management) Regulations 1994. The role of the client*, 2000

- HSE, CIS40, *The role of the planning supervisor*, 2000

- HSE, CIS41, *The role of the designer*, 1995

- HSE, CIS42, *The pre-tender health and safety plan*, 1995

- HSE, CIS43, *The health and safety plan during the construction phase*, 1995

- HSE, CIS44, *The health and safety file*, 1995.

- HSE, INDG183, *5 steps to risk assessment*, 1998 (ISBN 0 7176 1580 4).

- HSE, INDG220, *A guide to The Construction (Health, Safety and Welfare) Regulations 1996*, 1996 (ISBN 0 7176 1161 2).

- HSE, INDG275, *Managing health and safety. Five steps to success*, 1998.

- HSE, PML54REV, *CDM Regulations – how the Regulations affect you*, 1996.

Chapter 3 Getting ready

- CITB, FTR 007C, *Safe Start Supervisor's Safety Check Card and Training Notes.*

Chapter 4 Personal protective equipment

- HSC, L25, Personal protective equipment at work. *Guidance on Personal Protective Equipment at Work Regulations 1992*, 1992 (ISBN 0 7176 0415 2).

- HSE, INDG174, *A short guide to the Personal Protective Equipment at Work Regulations 1992*, 1995 (ISBN 0 7176 0889 1).

- HSE, L102, *Construction (Head Protection) Regulations 1989. Guidance on Regulations*, 1998 (ISBN 0 7176 1478 6).

Chapter 5 Access/egress

- CIRIA, SP121, *Temporary access to the workface – a handbook for young professionals*, 1995 (ISBN 0 86017 422 0).

Chapter 6 Bottled gases

☐ HSE, CIS11, *Safe use of propane and other LPG cylinders,* 1988.

☐ HSE, CS6, *The storage and use of LPG on construction sites,* 1981.

• LPG Association, *Code of Practice 7 – Storage of Full and Empty LPG Cylinders and Cartridges,* 1998.

Chapter 7 Building maintenance

• HSE, HSG58, *Evaluation and inspection of buildings and structures,* 1990 (ISBN 0 1188 5441 0).

Chapter 8 Care with asbestos

☐ BSI, EN 45001:1989, *General criteria for the operation of testing laboratories,* 1989.

• HSC, L27, *The Control of Asbestos at Work Regulations 1987. Approved code of practice,* 3rd edn, 1999 (ISBN 0 7176 1673 8).

• HSC, L28, *Work with asbestos insulation, asbestos coating and asbestos insulating board. Control of Asbestos at Work Regulations 1987. Approved code of practice,* 3rd edn, 1999 (ISBN 0 7176 1674 6).

• HSE, EH10, *Asbestos – exposure limits and measurement of airborne dust concentrations,* 1995 (ISBN 0 7176 0907 3).

• HSE, EH47, *Provision, use and maintenance of hygiene facilities for work with asbestos insulation and coatings,* 1990 (ISBN 0 1188 5567 0).

• HSE, EH51, *Enclosures provided for work with asbestos insulation, coatings and insulating board,* 1999 (ISBN 0 7176 1700 9).

• HSE, INDG188, *Asbestos alert for building maintenance, repair and refurbishment workers,* 2000 (ISBN 0 7176 1209 0).

• HSE, HSG189/1, *Controlled asbestos stripping techniques for work requiring a licence,* 2nd edn, 1999 (ISBN 0 7176 1666 5).

• HSE, HSG189/2, *Working with asbestos cement,* 1999 (ISBN 0 7176 1667 3).

• HSE, INDG223REV1, *Managing asbestos in workplace buildings,* 1996 (ISBN 0 7176 1179 5).

• HSE, INDG289, *Working with asbestos in buildings,* 1999 (ISBN 0 7176 1697 5).

35.5 Bibliography

Chapter 9 Chemicals, dust and fumes

- CIRIA, R125, *A guide to the control of substances hazardous to health in design and construction,* 1993 (ISBN 0 86017 371 2).
- HMSO, SI 1993/1746, *The Chemicals (Hazard Information and Packaging) Regulations 1993,* 1993 (ISBN 0 11 034746 3) and subsequent amendments.
- HSE, CIS26REV, *Cement,* 1996.
- HSE, EH40, *Occupational exposure limits 2000,* 2000 (ISBN 0 7176 1730 0).
- HSE, HSG97, *A step by step guide to COSHH assessment,* 1999 (ISBN 0 7176 1446 8).
- HSE, INDG181REV1, *The complete idiot's guide to CHIP,* 1999 (0 7176 2439 0).
- HSE, INDG182, *Why do I need a safety data sheet?,* 1994 (ISBN 0 7176 0895 6).
- HSE, MSB9, *Save your skin,* 1990.
- Stationery Office, SI 2000/2381, *The Chemicals (Hazard Information and Packaging Supply)(Amendment) Regulations 2000,* 2000 (ISBN 0 11 099805 7).

Chapter 10 Confined spaces

- CFL Vision, *Watch that Space – Construction* (video).
- HSE, CIS15, *Confined spaces,* 1991.
- HSE, INDG258, *Safe work in confined spaces,* 1997 (ISBN 0 7176 1442 5).

Chapter 11 Cranes and hoists

- BSI, BS 7121, *Code of Practice for Safe Use of Cranes,* Parts 1 (1989), 2 (1991), 3 (2000), 4 (1997), 5 (1997), 11 (1998) and 12 (1999).
- CIRIA, SP121, *Temporary access to the workface – a handbook for young professionals,* 1995 (ISBN 0 86017 422 0).
- CIRIA, SP131, *Crane stability on site: an introductory guide,* 1996 (ISBN 0 86017 456 5).
- CITB, CJ 502, *Mobile Crane Operator's Safety Guide,* 1979.
- HSE, CIS13, *Construction goods hoists,* 1988.

▫ HSE, CIS19, *Safe use of mobile cranes on construction sites,* 1991
● HSE, L64, *Safety signs and signals. The Health and Safety (Safety Signs and Signals) Regulations 1996. Guidance on Regulations,* 1996 (ISBN 0 7176 0870 0).
● Stationery Office, SI 1998/2307, *The Lifting Operations and Lifting Equipment Regulations 1998,* 1998 (ISBN 0 11 079598 9).

Chapter 12 Demolition
● CIRIA, SP121, *Temporary access to the workface – a handbook for young professionals, 1995* (ISBN 0 86017 422 0).
● HSE, CIS45, *Establishing exclusion zones when using explosives in demolition,* 1995.
● HSE, HSG150, *Health and safety in construction,* 1996 (ISBN 0 7176 1143 4).

Chapter 13 Electricity
▫ HSE, CIS6, *Portable electric tools and equipment,* 1988.
● HSE, GS6REV, *Avoidance of danger from overhead electric power lines,* 1997 (ISBN 0 7176 1348 8).
● HSE, HSG107, *Maintaining portable and transportable electrical equipment,* 1994 (ISBN 0 7176 0715 1).
● HSE, HSG141, *Electrical safety on construction sites,* 1995 (0 7176 1000 4).
● Yorkshire Electricity, *Prevention of Accidents: Building and Construction,* 1996.

Chapter 14 Environment
● CIRIA, *Environmental handbook for building and civil engineering projects,* 2000. C512, *Part 1: design and specification* (ISBN 0 86017 512 X); C528, *Part 2: construction phase* (ISBN 0 86017 528 6); C529, *Part 3: demolition and site clearance* (ISBN 0 86017 529 4).
● CIRIA, SP96, *Environmental assessment,* 1994 (ISBN 0 86017 379 8).
● CIRIA, SP122, *Waste minimisation and recycling in construction – a review,* 1995 (ISBN 0 86017 428 X).
● CIRIA, SP141V, *Building a cleaner future* (video), 1996.

35.7 Bibliography

- HMSO, *Environmental Protection Act 1990,* 1990 (ISBN 0 10 544390 5).
- HMSO, *Waste Management – The Duty of Care. A Code of Practice,* 1996 (ISBN 0 11 753210 X).
- HMSO, *Water Resources Act 1991,* 1991 (ISBN 0 10 545791 4).
- HMSO, SI 1988/819, *The Collection and Disposal of Waste Regulations 1988,* 1988 (ISBN 0 11 086819 6).
- HMSO, SI 1989/1790, *The Noise at Work Regulations 1989,* 1989 (ISBN 0 11 097790 4).
- HMSO, SI 1992/339, *The Trade Effluents (Prescribed Processes and Substances) Regulations 1992,* 1992 (ISBN 0 11 023339 5).
- HMSO, SI 1992/588, *The Controlled Waste Regulations 1992,* 1992 (ISBN 0 11 023588 6).
- Stationery Office, *Environment Act 1995,* 1995 (ISBN 0 10 542595 8)
- Stationery Office, SI 1999/437, *The Control of Substances Hazardous to Health Regulations 1999,* 1999 (ISBN 0 11 082087 8).

Chapter 15 Excavation

- CIRIA, R97, *Trenching practice,* 2nd edn, 1992 (ISBN 0 86017 192 2).
- CIRIA, TN95, *Proprietary trench support systems,* 3rd edn, 1986 (ISBN 0 86017 264 3).
- HSE, HSG185, *Health and safety in excavations – be safe and shore,* 1999 (ISBN 0 7176 1563 4).

Chapter 16 Falling

- HSC, HSG33, *Health and safety in roof work,* 1998 (ISBN 0 7176 1425 5).
- HSE, INDG284, *Working on roofs,* 1999.
- HSE, *Deadly maintenance. Roofs – a study of fatal accidents at work,* 1985.

Chapter 17 Fire

- Building Employers Confederation, *Fire Prevention on Construction Sites,* 4th edn, 1997.
- HSE, CIS51, *Construction fire safety,* 1997.

- HSE, HSE8REV, *Take care with oxygen. Fire and explosion hazards in the use and misuse of oxygen,* 2nd edn, 1999 (ISBN 0 7176 2474 9).

- HSE, HSG140, *The safe use and handling of flammable liquids,* 1996 (ISBN 0 7176 0967 7).

- HSE, HSG168, *Fire safety in construction work,* 1997 (ISBN 0 7176 1332 1).

- HSE, INDG227, *Safe working with flammable substances,* 1996 (ISBN 0 7176 1154 X).

Chapter 18 Frame erection

- BCSA, 20/89, *Structural Steelwork – Erection,* 1989 (ISBN 0 85073 021 X).

- CIRIA, SP121, *Temporary access to the workface – a handbook for young professionals,* 1995 (ISBN 0 86017 422 0).

- HSE, GS28, *Safe erection of structures* (in four parts), 1984–1986 (ISBN 0 1188 3530 0).

Chapter 19 Manual handling

- HSC, CIS37, *Handling heavy building blocks,* 1999.

- HSC, L23, *Manual handling. Manual Handling Operations Regulations 1992. Guidance on Regulations,* 2nd edn, 1998 (ISBN 0 7176 2415 3).

- HSE, AS23, *Manual handling solutions for farms,* 2000.

- HSE, HSG115, *Manual handling. Solutions you can handle,* 1994 (ISBN 0 7176 0693 7).

- HSE, INDG143REV1, *Getting to grips with manual handling. A short guide for employers,* 2000.

Chapter 20 Noise

- IBSI, BS 5228-1:1997, *Noise and vibration control on construction and open sites. Code of practice for basic information and procedures for noise and vibration control,* 1997.

- CIRIA, PR70, *How much noise do you make? A guide to assessing and managing noise on construction sites,* 1999 (ISBN 0 86017 870 6)

- CIRIA, TN138, *Planning to reduce noise exposure in construction,* 1990 (ISBN 0 86017 317 8)

35.9 Bibliography

- HSE, CRR73, *Dust and noise in the construction process,* 1995 (ISBN 0 7176 0768 2).
- HSE, INDG99, *Noise at work. Advice to employees,* 1998 (ISBN 0 7176 0962 6).
- HSE, INDG127REV, *Noise in construction. Further guidance on the Noise at Work Regulations 1989,* 1995.
- HSE, Noise Guides 1/2 and 3–8, *Noise at work,* 1989.

Chapter 21 Pressure testing

- BSI, BS 8005-5:1990, *Sewerage. Guide to rehabilitation of sewers,* 1990.
- BSI, BS 8301:1985, *Code of practice for building drainage,* 1985.
- HSE, GS4, *Safety in pressure testing,* 3rd edn, 1998 (ISBN 0 7176 1629 0).

Chapter 22 Public safety and site security

HSE, HSG151, *Protecting the public – your next move,* 1997 (ISBN 0 7176 1148 5).

Chapter 23 Scaffolding

BSI, BS 5973:1993, *Code of practice for access and working scaffolds and special scaffold structures in steel,* 1993.
- CIRIA SP121, *Temporary access to the workface – a handbook for young professionals,* 1995 (ISBN 0 86017 422 0).
- CITB, CE 509, *A Guide to Practical Scaffolding,* 1987.
- HSE CIS10REV, *Tower scaffolds,* 3rd edn, 1997.
- HSE CIS49, *General access scaffolds and ladders,* 1997.
- HSE GS15, *General access scaffolds,* 1982.
- HSE GS42, *Tower scaffolds,* 1987.

Chapter 24 Site investigation

- British Drilling Association, *Guidance Notes for the Safe Drilling of Landfills and Contaminated Land,* 1992.
- HMSO, SI 1996/1592, *The Construction (Health, Safety and Welfare) Regulations 1996,* 1996 (ISBN 0 11 035904 6).
- Scott Wilson Kirkpatrick, *Health and Safety Manual,* Part 4, Section 5, 1994.

Chapter 26 Small plant and equipment

- Forestry Commission, Safety Guides 301–310.
- Stationery Office, SI 1998/2306. *The Provision and Use of Work Equipment Regulations 1998*, 1998 (ISBN 0 11 079599 7).

Chapter 27 Temporary works

- HSE, *Final Report of the Advisory Committee on Falsework*, 1976.
- HSE, HSG32, *Safety in falsework for in-situ beams and slabs*, 1987 (ISBN 0 1188 3900 4).

Chapter 28 Vehicles and site transport

- HSE, HSG136, *Workplace transport safety. Guidance for employers*, 1995 (ISBN 0 7176 0935 9).
- HSE, INDG31, *Danger! Transport at work*, 1985.
- HSE, INDG148, *Reversing vehicles*, 1993 (ISBN 0 7176 1063 2).

Chapter 29 Working on live roads

HMSO, *New Roads and Street Works Act 1991*, 1991 (ISBN 0 10 542291 6).

- Stationery Office, *Safety at Street Works and Road Works. A Code of Practice*, 2001 (ISBN 0 11 551958 0).
- Stationery Office, *Traffic Signs Manual. Chapter 8, Traffic safety measures and signs for road works and temporary situations*, Volumes 1 and 2, 1991 (ISBN 0 11 550937 2).

Chapter 30 Working over water

- CIRIA, SP137, *Site safety for the water industry*, 1997 (ISBN 0 86017 460 3).

Chapter 31 Working alone

- HSE, INDG73REV, *Working alone in safety. Controlling the risks of solitary work*, 1998 (ISBN 0 7176 1507 3).
- RICS, *Surveying Safely*, 1991.

35.11 Bibliography

Chapter 33 Dealing with accidents

- HSC, L74, *First aid at work. The Health and Safety (First Aid) Regulations 1981. Approved code of practice and guidance,* 1997 (ISBN 0 7176 1050 0).

- HSE, HSE31REV1, *RIDDOR explained. The Reporting of Injuries, Diseases and Dangerous Occurrences Regulations,* 1999 (ISBN 0 7176 2441 2)

- HSE, L73, *A Guide to The Reporting of Injuries, Diseases and Dangerous Occurrences Regulations* 1995, 1999 (ISBN 0 7176 2431 5).

Chapter 34 Accident investigation

- HSE, INDG113, *Your firm's injury rates and how to use them,* 1991.

Information

For health and safety information and advice, contact the HSE public enquiry point:

HSE Information Centre
Broad Lane
Sheffield S3 7HQ
Tel: 0114 289 2345
Fax: 0114 289 2333

website: www.hse.gov.uk/contact/

Publications

All HSE publications are available from:

HSE Books
PO Box 1999
Sudbury
Suffolk CO10 6FS
Tel: 01787 881165
Fax: 01787 313995.

website: www.hsebooks.co.uk

All HSE current publications are listed in Publication In Series: List of HSC/HSE Publications, which is updated regularly and is available free from HSE Books.

Report forms

Until early 1992 HMSO published forms for recording the results of examinations and tests of lifting equipment and plant and for specifying safe working loads. These forms summarised the information prescribed under the various Regulations in a convenient format.

The forms are no longer available, but where companies have stocks of the original they may still be used. Copies of forms F54, 75, 80, 87, 88, 91, 96, 97 and 2040 may still be made and used.

Subject index

Drowning, 30.2
Dust, 4.4, 9.2, 9.4, 12.4

E
Ear defenders, 4.3, 20.2
Edge protection, 15.4, 16.3
Electrical hazards,
7.2, 13.1, 13.2, 31.1
Electricity, 13
Electricity at Work Regulations, 1.3
Emergency services and procedures,
2.3,17.2, 17.4, 25.2, 30.2, 32.1
Environment, 14
Environment Agency, 14.2
Environmental assessment, 14.2
Environmental Protection Act, 14.1
Erection sequence, 18.2, 18.4
Excavations, 15
Excavating, 1.6, 2.3
Explosives, Starting point, 6.1
Extension cables, 13.2
Eye protection, 4.3

F
Face masks, 4.4
Fall arresters, 4.4
Falling, 2.3, 16
Falls, 12.2
Falsework, 27.1
Fencing, 22.2
Fillers, 9.4
Fire, 2.3, 17
Fire extinguishers, 6.2, 17.3
First aid, 25.2, 31.2, 33.1
Flammable:
atmosphere, 10.2
gases, 17.3, 21.2
liquids, 17.1
substances, 9.3
Flammable materials stores,
17.3, 17.4
Flat roof, 16.4
Foot protection, 4.2
Formwork, 27.1

Fragile roofs, 16.4
Frame erecting, 1.6, 18.2-5
Frame erection, 18
Frost, 15.2
Fumes, 9.2, 9.4
Fuses, 13.2

G
Gangways, 5.2
Gases, 6.1, 10.2, 17.3, 21.1, 21.2
Getting ready, 3
Glues, 9.4
Ground contamination, 9.4

H
Hand protection, 4.4
Harmful substances, 9.3
Hazard symbols, 9.3, 9.5
Hazards, 9.2, 14.1, 32.1
Head protection, 4.2
Health and Safety at Work etc Act,
1.2, 1.3, 22.2
Health and Safety Commission, 1.2
Health and Safety Executive,
1.2, 36.1
Health surveillance, 9.4
Hearing protection, 4.3
Hoists, 11.5, 11.6
Hot work, 1.6, 17.4
HSE publications and
public information points, 36
Hydraulic testing, 21.2
Hygiene facilities, 25.1, 25.2, 25.3

I
Illness, 3.2
Industrial deaths, 1.1
Inflammable – see Flammable
Inspection, 2.3
Ionising radiation, Starting point
Irritant substances, 9.3

Subject Index

Subject Index

Subject Index

Weedkiller, 9.4
Welding, 17.4
Welfare facilities, 2.3, 25.1
Working:
    alone, 3.2, 31.1
    in confined spaces, 10.2
    over water,
        1.6, 30.1, 30.2, 31.1
    platform, 23.6
Working Alone, 31
Working on Live roads, 29
Working over Water, 30

X, Y, Z
Your actions, 1.7, 32.1
Your responsibilities, 1